HYDROFRACKING

WHAT EVERYONE NEEDS TO KNOW ®

HYDROFRACKING

WHAT EVERYONE NEEDS TO KNOW®

ALEX PRUD'HOMME

OXFORD
UNIVERSITY PRESS

OXFORD
UNIVERSITY PRESS

Oxford University Press is a department of the University of Oxford.
It furthers the University's objective of excellence in research, scholarship,
and education by publishing worldwide.

Oxford   New York
Auckland   Cape Town   Dar es Salaam   Hong Kong   Karachi
Kuala Lumpur   Madrid   Melbourne   Mexico City   Nairobi
New Delhi   Shanghai   Taipei   Toronto

With offices in
Argentina   Austria   Brazil   Chile   Czech Republic   France   Greece
Guatemala   Hungary   Italy   Japan   Poland   Portugal   Singapore
South Korea   Switzerland   Thailand   Turkey   Ukraine   Vietnam

Oxford is a registered trademark of Oxford University Press
in the UK and certain other countries.

"What Everyone Needs to Know" is a registered trademark of Oxford
University Press.

Published in the United States of America by
Oxford University Press
198 Madison Avenue, New York, NY 10016

Library of Congress Cataloging-in-Publication Data
Prud'homme, Alex.
Hydrofracking / Alex Prud'homme.
pages cm—(What everyone needs to know)
Includes bibliographical references and index.
ISBN 978-0-19-931125-5 (pbk. : alk. paper)—ISBN 978-0-19-931126-2
(cloth : alk. paper)   1. Hydraulic fracturing—Popular works.
2. Shale gas reservoirs—Popular works.   3. Oil wells—
Hydraulic fracturing—Popular works.   I. Title.
TN880.2.P78 2013
622'.3383—dc23
2013028962

CONTENTS

GLOSSARY

acid. A generic term used to describe a treatment fluid typically comprising hydrochloric acid and a blend of acid additives.

annulus. The space between a casing string and the borehole or between two casing strings.

aquifer. A water-bearing stratum of permeable rock, sand, or gravel.

biocide. An additive that eliminates bacteria in the water that produce corrosive byproducts.

blender. The equipment used to prepare the slurries and gels commonly used in fracture stimulation treatments.

borehole. A hole drilled into the earth by people in search of natural gas and oil.

breaker. An additive that reduces the viscosity of fluids by breaking long-chain molecules into shorter segments.

CAS. Chemical Abstract Service, a division of the American Chemical Society, whose objective is to find, collect, and organize publicly disclosed substance information.

CAS number. A unique number assigned by the CAS that identifies a chemical substance or molecular structure.

The List compiled from FracFocus: http://fracfocus.org/glossary of terms, and the State of California, Department of Conservation website: http://www.conservation.ca.gov/index/Pages/glossary-frk.aspx

casing. Pipe placed in an oil or gas well to (1) prevent the wall of the hole from caving in; (2) prevent movement of fluids from one geologic formation to another; (3) provide a means of maintaining control of formation fluids and pressure as the well is drilled.

casing string. Pipe that lines a well after it has been drilled. It is formed from sections of tube that have been screwed together.

collapse strength of casing. The pressure necessary to collapse a well casing, tubing, or drill pipe inside a well. The collapse strength of casing can be calculated from the yield strength of the metal and diameter of and wall thickness of the casing, tubing, or drill pipe.

conductor casing. Generally, the first string of casing in a well. It may be lowered into a hole drilled into the formations near the surface and cemented in place, or it may be driven into the ground by a special pile driver. Its purpose is to prevent the soft formations near the surface from caving in and to conduct drilling mud from the bottom of the hole to the surface when drilling starts. (Also known as "conductor pipe" and "drive pipe.")

casing shoe. A tapered, bullet-nosed piece of equipment often found on the bottom of a casing string. The device guides the casing toward the center of the borehole and minimizes problems associated with hitting rock ledges or washouts as the casing is lowered into the well. (Also known as a "guide shoe.")

cement. A mixture of sand, water, and a binding agent with no aggregates.

Cement Bond Log. A geophysical log that graphically displays the bond between cement and casing.

clay stabilizer. An additive that prevents clays from swelling or shifting.

conductor casing. The first casing string placed in a borehole. The purpose of the conductor is to prevent the collapse of the hole in unconsolidated material, such as soil.

corrosion inhibitor. An additive used in acid treatments to prevent corrosion of pipe by the corrosive treating fluid.

cross-linker. An additive that reacts with multiple-strand polymers to couple molecules, creating a fluid of high but closely controlled viscosity.

Darcy's Law. The mathematical equation that quantifies the ability of fluid to flow through porous material such as rock.

data van. The truck used to monitor all aspects of the hydraulic fracturing job.

DOE. The US Department of Energy.

drilling rig. The equipment used to drill the borehole.

EPA. The US Environmental Protection Agency.

EPCRA. The Emergency Planning and Community Right to Know Act. In 1986, Congress enacted EPCRA, a statute that established requirements for federal, state, and local governments, tribes, and industry regarding emergency planning and community right-to-know" reporting on hazardous and toxic chemicals. One provision of EPCRA remains highly controversial. As the EPA notes, "All information submitted pursuant to EPCRA regulations is publicly accessible, unless protected by a trade secret claim."

frac tank. The container used to store water or proppant that will be used for hydraulic fracturing.

friction reducer. An additive used to reduce the friction forces on tools and pipes in the borehole.

gelling agent. An additive that increases the viscosity of a fluid without substantially modifying its other properties.

groundwater. Water in a saturated zone under the earth's surface.

GWPC. Ground Water Protection Council.

hydrology. The study of the flow of water.

intermediate casing. Provides protection against caving in of weak or abnormally pressured formations and enables the use of drilling fluids used to drill into lower formations.

intermediate casing string. The string of casing set in a well (after the surface casing but before production casing is set) to keep the hole from caving in and to seal off formations. In deep wells, one or more intermediate strings may be required.

IOGCC. Interstate Oil and Gas Compact Commission.

MSDS. Material Safety Data Sheet.

mechanical integrity. The measure of a well's casing, tubing, packer, and cement to contain fluids traveling up and down the well without the fluids leaking into surrounding geologic formations.

natural gas. Methane CH_4 (with or without impurities such as nitrogen). Natural gas is often classified as either biogenic (of biological origin) or thermogenic (of thermal or heat origin).

NGWA. National Ground Water Association.

nonfreshwater fluids. Water with total dissolved solids (TDS) of greater than 3,000 parts per million (ppm) or any other fluid used in oil and gas production, including hydraulic fracturing fluids.

other cement evaluation method. Used to determine the quality of the cement bond between a well's casing and the geologic formations surrounding a borehole.

oxygen scavenger. An additive that prevents corrosion of pipes by oxygen.

packer. A downhole device used in completions to isolate the casing-tubing annulus from the production conduit, enabling controlled production, injection, or treatment.

perforate. To pierce holes in the casing and cement in a well to allow formation fluids, such as oil and gas, to enter into a well and to allow fluids to be injected into a geologic formation. Perforating is accomplished using a perforating gun, or perforator.

perforated interval. A section of casing that has been perforated during hydraulic fracturing.

permeability. The quantification of how easily fluids, such as oil, gas, and water, flow through the pore spaces in a geologic formation and into the borehole. Rocks have vertical, horizontal, and tangential permeability.

pH adjusting agent. An additive that adjusts the acidity/alkalinity balance of a fluid.

Poisson's ratio. A mechanical property that can be used to predict the direction in which fractures will occur in a given geologic formation.

When a geologic formation is compressed in one direction, it tends to expand in the other two directions perpendicular to the direction of compression. This phenomenon is called the Poisson effect.

production casing. The casing string set near the bottom of a completed borehole through which oil or natural gas is produced.

proppant. A granular substance (sand grains, crushed walnut shells, aluminum pellets, or other material) that is carried in suspension by the fracturing fluid. Proppant keeps fractures open in a formation when fracturing fluid is withdrawn after a fracture treatment.

radial cement evaluation log. A continuous log (graph) created by running a sonic transmitter and receiver down a well to determine the quality of the cement bond between the casing and the geologic formations surrounding a well.

reservoir. A bed of rock containing oil or natural gas.

reverse engineering. The reproduction of a product, through analysis of its structure, function, and operation.

saturated zone. The subsurface zone where the interstitial spaces of rock are filled with water.

shale. A fine-grained sedimentary rock that may contain oil or natural gas but which may not be producible naturally.

site. The location of a well including the area used for fluid storage and well treatment.

spring. The intersection of groundwater and surface water.

surface casing. The casing string set below freshwater aquifers to prevent their contamination.

surfactant. A chemical that acts as a surface active agent. This term encompasses a multitude of materials that function as emulsifiers, dispersants, oil-wetters, water-wetters, foamers, and defoamers.

total dissolved solids (TDS). The total amount of solids, such as minerals, salts, or metals, that are dissolved in a given volume of water.

toxicology. The study of symptoms, mechanisms, treatment, and detection of poisoning.

TRI. Toxic Release Inventory.

true vertical depth. The vertical distance from a point in the well (usually the current or final depth) to the surface.

tubing. A small-diameter pipe that is run inside well casing to serve as a conduit for the passage of oil and gas to the surface. Tubing can be a permanent or temporary part of the borehole.

tubing strings. The entire length of tubing in a well.

unsaturated zone. The subsurface zone where the interstitial spaces of rock contain but are not completely filled with water.

vadose zone. The subsurface zone between the surface and the unsaturated zone through which water travels.

Variable Density Log. The geophysical log that is a graphic representation of the bond between the cement and the borehole.

well. The hole made by the drilling bit, which can be open, cased, or both. Also called borehole, hole, or wellbore.

wellbore. A borehole; the hole drilled by a drill bit; also called a borehole or hole.

well stimulation. Any of several operations used to increase production by increasing the permeability of an oil- or gas-bearing formation, such as acidizing or hydraulic fracturing.

Young's modulus. Used in drilling, it is a measure of the stiffness of elastic of a geologic formation, defined as the ratio of the stress along an axis over the strain along that axis. Young's modulus can help determine how wide fractures are likely to be in a formation that will be hydraulically fractured.

PREFACE: WHY I WROTE THIS BOOK

I was first confronted by the intense emotions around hydro-fracking at a public meeting in New York City in November 2009. It was a cold, blustery night in downtown Manhattan, but over a thousand people streamed into a high school auditorium to learn about the potential benefits and haz-ards of extracting natural gas from in and around the city's upstate watershed. I was there to research my book, *The Ripple Effect: The Fate of Freshwater in the Twenty-First Century*, and was curious to know what impact hydrofracking might have on the quality and quantity of the drinking water supplied to over nine million people every day.

The debate that night centered on the Marcellus Shale, which is a 95,000-square-mile swath of gas-rich rock that underlies parts of five states: New York, Pennsylvania, Maryland, West Virginia, and eastern Ohio. The stakes in play there—finan-cial, environmental, political, and social—are enormous. The Marcellus deposit is thought to be the single largest energy deposit in the United States, and the second-largest gas deposit in the world (after the South Pars/North Dome gas-field, shared by Qatar and Iran). The Marcellus is estimated to contain at least 500 trillion cubic feet of natural gas, which is enough to power all American homes for 50 years.[1]

Shale is a dense layer of sedimentary rock that lies a mile or more underground in deposits sprinkled across the country (and, indeed, around the world). Natural gas or oil trapped in shale formations is known as "shale gas" or "shale oil," and is chemically identical to gas and oil taken from traditional wells. Geologists have known about shale reserves for years, but until recently they have been too difficult to access. In the last decade, however, industry and government groups have pushed a technology called hydraulic fracturing, or "fracking," which has unlocked enormous quantities of shale oil and gas and set off an environmentalist backlash.

Hydrofracking has created jobs, spurred industry, lowered carbon emissions, and provided an economic boon to many communities across the country. (While there is great interest in the technology worldwide, hydrofracking has been commercialized only in North America thus far.) Yet, while the temptations of "fracked" energy are great, critics say that it pollutes the water, ground, and air, and that these costs outweigh its benefits.

It was against this backdrop that I attended the meeting in New York City in 2009. The large auditorium was packed to standing-room only that night. Roughly a quarter of the crowd supported hydrofracking; another quarter had not made up their minds; and the remaining half were opposed. Some attendees wore suits or high heels, some came in camouflage and blue jeans, others were dressed up as mountains, fish, or rivers. Red-faced politicians stirred the crowd with fiery rhetoric; state regulators and energy executives kept a low profile; journalists swirled around the auditorium; and citizens asked pointed questions.

When gas companies began to explore rural upstate New York in the early 2000s, many residents leased their property for modest fees. Some were paid as little as $3 per acre plus a 12.5 percent royalty; by 2007 lease prices averaged about $25 an acre, plus royalties of 12.5 percent; by 2009, prices had skyrocketed to $6,000 an acre, plus royalties of 20 percent.[2] The region was mired in an economic slump, and many residents and businesses were pushing then-governor David Paterson to open state-owned land to hydrofracking to generate jobs and revenue. But Michael Bloomberg, mayor of New York, cautioned that fracking "is not a risk that I think we should run."[3] Since that night in 2009, the two sides have only become more polarized. In New York State, for example, the public remains almost evenly split on the issue, with 39 percent in favor of hydrofracking and 43 percent opposed to it, according to a 2013 Siena poll.[4] The *Wall Street Journal* opines, "fracking could be the difference between economic life and death" for New York.[5] But celebrity opponents, like Yoko Ono and the actor Mark Ruffalo (who lives in upstate New York), shoot back, "You can't say that we have climate change and we have to fight it, and then... say we're going to move forward with hydrofracking.... You can't have both."[6] And "Fracktivists" note that if fracking fluids—some of which are toxic or carcinogenic (as is benzene)—pollute the city's carefully protected watershed, New York will be forced by EPA regulations to build a $10 billion filtration plant that will cost taxpayers millions of dollars a year to operate.[7]

Current New York governor Andrew Cuomo has no easy answers. Hydrofracking advocates, such as the Joint Landowners Coalition of New York, have pressured state legislators to approve the process for nearly five years, and complain that Cuomo is stalling.[8] But opposition groups threaten to label Cuomo—who is said to have presidential ambitions—a traitor and sellout if he allows hydrofracking.[9] Cuomo was given a momentary reprieve in early 2013. The day before state regulators were set to issue an environmental impact statement, the state Department of Health requested more time to review three new studies. The state assembly imposed a two-year moratorium on new hydrofracked wells to await results of the studies.[10]

The argument in New York mirrors the national dispute over hydrofracking, and foreshadows a worldwide debate as shale gas and oil become increasingly important to the global energy equation.

In the meantime, my own position on hydrofracking continues to evolve. In 2009, informed by my research into water-related issues, I was opposed to hydrofracking. There is little question that the process is inherently risky: it uses huge volumes of water and has set off local "water wars" in arid states such as Colorado, Texas, and California. Moreover, shale wells can pollute the air and groundwater. Once hydrofracked, each well generates millions of gallons of toxic wastewater, which includes secret chemical mixtures and naturally occurring radioactive elements that are difficult to clean and sequester. As hydrofracking technology spreads around the world, these challenges will become exponentially more difficult.

Yet the technology, practice, and oversight of hydrofracking have advanced since 2009, and it has become difficult to ignore the benefits of shale fuels. The scientific consensus holds that natural gas burns more cleanly than coal or oil, and thus reduces greenhouse gases; the economic consensus holds that hydrofracking creates jobs, revenue, and new supplies of energy; and the political consensus holds that natural gas is an effective "bridge fuel" to tide us over until renewable energy sources—such as wind, solar, geothermal, and hydropower—have been commercialized.

To put it bluntly, hydrofracking is neither all good nor all bad. Rather, it is a timely and important subject rendered in shades of gray. And it is one that is worth talking about and, indeed, arguing over. My aim in writing this book is to help spur a healthy, informed dialogue about an energy supply that we still have much to learn about and that is changing the world we live in.

HYDROFRACKING

WHAT EVERYONE NEEDS TO KNOW ®

INTRODUCTION: A TWENTY-FIRST-CENTURY BONANZA

In the second decade of the twenty-first century, America has found a new source of fuel. It is affordable, burns cleaner than coal and oil, and there is so much of it that some believe that supplies will last for over 100 years. What is this seemingly miraculous substance? It is called shale gas—natural gas trapped in shale rock "formations" (deposits) scattered across the country. When you hear about the "Marcellus Shale" or the "Bakken Shale," these formations are what people are referring to. Some of them contain crude oil, but it is the gas that is the focus of attention.

Until recently, these shale formations—made of dense sedimentary rock, often buried a mile underground—have been considered too expensive to prospect. As demand for energy has spiked, and the nation worries about jobs, energy independence, and the impact of greenhouse gases, industry and government groups have aggressively sought new sources of power, and the question of how to extract shale gas has risen. In the last decade an established technology called hydraulic fracturing—also known as hydrofracking, hydrofracturing, or, most commonly, "fracking" (also spelled "fracing")—has seemed to offer the best way of getting at it. Retooled for the twenty-first century, hydrofracking has unlocked those reserves of shale oil and gas, and in the process set off both an energy revolution and an environmentalist backlash.

Hydrofracking has most famously been used to tap natural gas, and this book could have been titled *Shale Gas: What Everyone Needs to Know®*. But "hydrofracking" is a broader term that describes a process used to access both oil and gas, and for various other purposes (as I explain below), and it refers to a host of important energy issues that are being hotly debated.

Indeed, the word "hydrofracking" means different things to different people, but recently it has morphed into a catch-all term for two drilling techniques frequently used together: *horizontal drilling*, which allows energy companies to drill vertically and then laterally through long, narrow shale formations; and actual *hydraulic fracturing*, in which a "slurry"—a term I will deal with later—of fluids (water, sand, and chemicals—such as antifreeze, hydrochloric acid, and 2-BE ethylene glycol) is injected underground at very high pressure to crack open the dense shale rock, allowing gas or oil to flow to the surface, where it is captured for our use.

Fracking has proven effective, so much so that it has both radically transformed the energy landscape and led to an emotionally charged opposition from people concerned about its impact on human and environmental health. As you may have heard, hydrofracking is nothing if not controversial. Some commentators refer to the phenomenon as a "shale gas revolution" or a "new energy bonanza"; others refer to it as "an environmental nightmare."[1]

On the plus side, there are clear benefits to hydrofracking. The opening of new shale fields has created jobs, spurred industries, such as petrochemical and steel manufacturing, lowered carbon emissions (natural gas burns more efficiently than coal, and thus limits greenhouse gas emissions), and provided an economic boon to many communities, some of them among the poorest in the United States.[2] (While fracking has been tested in other countries, such as Great Britain and Holland, it is widely used only in the United States and Canada at the moment.)[3] Natural gas is an important fuel for power plants and industry, and it is increasingly used to fuel truck and bus fleets; at some point in the future, it is possible that gas supplied by hydrofracking could be used to power our cars. China and India are thought to have significant shale deposits, and as fracking technology spreads, it is likely that shale fuels will be used worldwide for many years.

Because natural gas emits less CO_2 than other fossil fuels and does not produce nuclear energy's radioactive waste, the French bank Société Générale has deemed it "the fuel of no choice."[4] And because fracking has unleashed a near tsunami of usable energy hydrocarbons, the International Energy Agency (IEA), a Paris-based consortium of energy interests, predicts that by 2035 the United States will have achieved a long-sought goal of energy independence, and be "all but self-sufficient in net terms."[5]

The United States has abundant natural gas reserves— over 2,203 trillion cubic feet of recoverable reserves, according to estimates by the US Energy Information Administration (EIA),[6] a division of the Department of Energy—which some believe could last for 100 years or more.[7] Yet so much gas has been fracked in the United States that the industry has become a victim of its own success: a glut dropped the price of natural gas from nearly $14 per million British Thermal Units (MBTU, also known as MMBTU) to about $4 per MBTU in 2012.[8] American energy companies are cutting back on shale gas production and looking to export gas and fracking technology. After decades of geopolitical tension over oil, the prospect of reducing America's dependence on the caprices of OPEC (the Organization of the Petroleum Exporting Countries) is welcome. Hydrofracking advocates boast that the tables have been turned at last, and soon America will become "the Saudi Arabia of gas."[9]

But critics charge that such progress comes at too steep a price: fracking pollutes the water, ground, and air and can be toxic to human and animal health. Reports of lax oversight have led to increased press scrutiny, demands for legislative action, and public alarm about the possibility of exploding wells, earthquakes, and climate-warming methane leaks. Opponents, from Pennsylvania to Poland, have swarmed legislators and blocked drilling sites. States like Vermont and New Jersey have joined Bulgaria and France in banning the practice outright. German beer brewers and individual towns in Colorado, Michigan, and England have called for a ban on the technology.

As the debate has intensified, hydrofracking has become the focus of political maneuvering and media attention. Energy companies have spent millions of dollars on pro-hydrofracking ads, lobbyists, and campaign donations to legislators. But the promotional effort has been dogged by an antifracking backlash that began as a grass-roots challenge and spread to national media and the halls of Congress. Lately, celebrities have adopted the antifrack message as a favorite cause. The documentary film *GasLand* (and *GasLand II*) and Matt Damon's feature film *Promised Land* drove the protest deep into popular culture. Yoko Ono and Sean Lennon paid for antifracking ads in the *New York Times* and wrote a protest song, "Don't Frack My Mother," to the tune of Bob Dylan's "The Times They Are a-Changin.'"[10] In a 2012 rant on *The Late Show*, David Letterman excoriated "the greedy oil and gas companies" and a compliant Environmental Protection Agency (EPA) for allowing fracking to "poison our drinking water" with toxic chemicals: "Ladies and gentlemen," he deadpanned, "we're screwed!"[11]

The bitter tone of the debate has been exacerbated by an absence of hard data and an excess of hyperbole on both sides. In writing this book I have striven for balance and parity, but it isn't always simple. For instance, some readers will note that chapter 5, "The Case for Hydrofracking," is shorter than chapter 6, "The Case against Hydrofracking." The reason for this is twofold: first, hydrofracking technology is still evolving, and much remains to be learned about the process and its impacts; second, the energy industry has refused to answer certain questions—such as the identity of possibly toxic chemicals in fracking fluids, which are shielded from public disclosure by trade-secret provisions.[12] The lack of information means that I can only describe what is known at this point, while a lack of industry answers leads people to speculate, make assumptions, and find fault.

In light of these developments, this is a timely moment for a dispassionate primer on hydrofracking. This book aims to explain what the process is, how it works, where and why it is used, and the pros and cons of fracking. I have endeavored to clear up misunderstandings and to promote an informed discussion about a process that is quickly transforming the way we use and think about energy. To accomplish this, the book is structured in three parts: Part I discusses fossil fuels and energy use in the twenty-first century, then explains what hydrofracking is, and how and where it is conducted; Part II covers the debate over hydrofracking, and details its pros and cons; Part III concludes with a look at the state of hydrofracking today, the major innovations on the way, and the promise of renewable energy in years to come.

Like all of the volumes in Oxford University Press's "What Everyone Needs to Know®" series, *Hydrofracking:What Everyone Needs to Know®* has been written for the lay reader in an accessible style. It is designed to be a useful introduction to a complex topic rather than a comprehensive academic treatise. In some cases, I have just touched on important technical issues, and for those interested in learning more I have provided a "Further Reading" section at the end of this book. To keep the material in these pages current, I have relied on important work done by many fine journalists, academics, industry publications, and government reports. My sources are listed in the notes at the end of this book (though hydrofracking is constantly evolving, and some of the information mentioned here will inevitably become outdated).

Hydraulic fracturing is a phenomenon with broad implications. For better or worse it is here to stay for the foreseeable future, and it behooves us to understand how it works, what are its rewards and dangers (or at least unknowns), and how to improve it going forward. Armed with this knowledge, readers will, I trust, draw their own conclusions about a technology that is redefining energy in the twenty-first century, and that we all therefore have a huge stake in.

PART I

HYDROFRACKING

WHAT, HOW, AND WHERE?

1

ENERGY IN CONTEXT: A FOSSIL FUEL PRIMER

What Are Fossil Fuels?

Buried deep in the earth are the remnants of earlier life forms—hundreds of millennia of rotting vegetation, decaying animals, and marine plankton. Today this "organic material" (so-called because it was once alive) has turned into rock that is laden with carbon, which can be burned for fuel. "Fossil fuels," such as coal, oil, and natural gas, were formed some 286 to 360 million years ago in the Carboniferous period, which predated the dinosaurs and is part of the Paleozoic Era.[1]

The word "carboniferous" is rooted in carbon, the basis of fossil fuels. During the Carboniferous period the earth was populated by trees, large leafy plants, and ferns, and its water bodies were rich with algae, a common phytoplankton. As these plants and animals died, they sank to the bottom of the swamps and oceans, forming layers of spongy earth called peat. Over millions of years, silt, sand, and clay covered the peat; as the layers of minerals built up, the peat was compressed until the water had been completely squeezed out of it. Eventually, the organic material was buried deeper and deeper and turned into rock. Increasing heat and geologic pressure transformed some of the rock into fossil fuels. Today, man drills deeply into the earth to reach the geologic formations in which coal, oil, and natural gas are trapped.

Unlike "renewable" wind or solar power, fossil fuels are "nonrenewable," or finite, energy supplies: once they are used, they are gone for good; their excavation and burning creates pollution, including greenhouse gases such as carbon dioxide (CO_2). Nevertheless, fossil fuels remain our primary source of energy.

That wasn't always the case. Until the late eighteenth century, the United States was an agrarian society, in which wind and water provided much of the power used to mill flour, saw wood, or irrigate crops. Burning wood or peat provided heat to people's homes. But as the nation industrialized, those fuels were largely replaced by fossil fuels—coal replaced wood, and eventually petroleum eclipsed coal—to power the steam engines that led to the Industrial Revolution. In the nineteenth century, animal oils, especially whale oil, used to produce light in oil lamps, were replaced by petroleum.

Today, the US Energy Information Administration (EIA) estimates that 25 percent of America's energy comes from petroleum, 22 percent from coal, and 22 percent from natural gas. Nuclear power accounts for 8.4 percent, and "renewable" energy—such as wind, solar, geothermal, or hydropower—supplies 8 percent.[2]

What Is Coal?

Coal is a combustible, brown or black sedimentary rock made of carbon, hydrogen, oxygen, nitrogen, and sulfur. Plentiful and cheap, it is the most abundant fossil fuel produced in the United States.[3] Coal is used around the world to fuel power plants and is an important resource for the iron and steel industries. But it is also highly polluting. According to the Union of Concerned Scientists, coal plants are the nation's primary source of CO_2, the leading greenhouse gas. A typical coal plant generates 3.5 million tons of CO_2 per year; in 2011, utility coal plants in the United States emitted 1.7 billion tons of CO_2.[4]

There are four major types of coal: anthracite, bituminous, sub-bituminous, and lignite. They are classified by the type and amount of carbon they contain, and by the amount of heat energy they produce:

- *Anthracite* is a scarce coal found only in Pennsylvania. It is the hardest coal, and has a high carbon content (86–97 percent) that provides more energy than other coals.
- *Bituminous* is the most abundant type of coal, containing 45–86 percent carbon, and is between 100 and 300 million years old. It is found in West Virginia, Kentucky, and Pennsylvania.
- *Sub-bituminous* coal is the second-most abundant type of coal in the United States. It contains 35–45 percent carbon, is over 100 million years old, and is mined in Wyoming.
- *Lignite* is the youngest and softest coal, and tends to be high in moisture. It contains only 25–35 percent carbon, but is high in hydrogen and oxygen. It is mainly used by power plants and is found in Texas and North Dakota.

Coal has been used as an energy source around the world for centuries. The Chinese, who thought coal was a stone that could burn, used it to smelt copper some 3,000 years ago. Coal is mined by various methods, including drilling of vertical and horizontal shafts; strip-mining, in which enormous shovels excavate surface layers of rock and earth to reveal coal seams; and mountaintop coal mining, in which entire mountaintops are removed, exposing coal deposits, while the waste rock is dumped into valleys and streams.

Coal is a cheap and plentiful fuel that has provided employment for millions of people around the world, and yet the carbon dioxide it produces is the most significant greenhouse gas, and it wafts ash into the air. This air pollution is environmentally destructive, and, according to the National Research Council, kills more than 10,000 Americans every year.[5]

Nevertheless, coal remains popular: even in this day of global networks and hybrid cars, coal provides two-fifths of the world's electricity.[6] In the last decade, the world's electricity production has doubled, and two-thirds of that increase has been powered by coal. Indeed, if this growth rate continues, coal could supplant oil as the world's primary energy source by 2017.[7]

The main drivers of increased coal use have been the surging economies of China and India, and steady demand from Western Europe. In 2001, according to the IEA, China's demand for coal was roughly equivalent to 60 million tons of oil.[8] By 2011, that demand had tripled, and China surpassed the United States as the largest energy producer in the world. (China's domestic coal industry produces more primary energy than oil from the Middle East does.) India's demand for coal is also booming. By 2017, the IEA estimates that India could overtake the United States as the world's second-largest user of coal.[9]

Even green-conscious Western Europe is burning more coal, although it is considered the most polluting form of energy. Coal is cheaper than natural gas in Europe, whose domestic gas industries lag far behind America's. Countries like Germany are pushing renewable energy, such as solar and wind power, which has eroded the creditworthiness of conventional utilities. US coal exports to European countries like Britain and Germany were up 26 percent in 2012 over 2011.[10] Coal consumption in economically battered Italy and Spain has also jumped, despite efforts to capitalize on Italy's wind and Spain's sunshine.

Thanks to the shale reserves tapped by hydrofracking, America is moving in the opposite direction. In 1988, coal provided 60 percent of US power; by 2012 that number had been cut in half, and coal generated only a third of US electricity—about the same amount as natural gas.[11] This dramatic shift was due in part to high costs and stricter emissions standards. Modern coal-fired plants take twice as long to build and are more expensive to run than gas-fired plants. And gas-fired plants meet environmental regulations far better than coal-fired plants.

Perhaps more important, the White House has been taking concrete steps to limit climate change at just the time that fracking has made shale gas affordable. The Obama administration touts itself as "pro energy,"[12] and has put rules in place to govern emissions of mercury and other airborne toxins by 2015. The Environmental Protection Agency (EPA) has proposed new limits on carbon emissions that would effectively ban new coal-fired plants after 2013 unless they are equipped with carbon capture and storage (or CCS) technology.[13] According to Navigant, a consulting group, expensive regulations and cheap gas could result in one-sixth of all coal-fired power plants in the United States—representing 50 gigawatts worth of power—being shuttered by 2017.[14] (A watt is a unit of power, equivalent to one joule per second, or 3.412 BTU/h. A gigawatt is equivalent to one billion watts.)

What Are Oil and Gasoline?

Oil is a fossil fuel that was created more than 300 million years ago, when diatoms died and decomposed on the sea floor. Diatoms are tiny sea creatures that convert sunlight directly into stored energy. After falling to the bottom of the ocean, they were buried under sediment and rock; the rock compressed the diatoms, trapping the energy in their pinhead-sized bodies. Subjected to great heat and pressure, the carbon eventually turned into liquid hydrocarbons (an organic chemical compound of hydrogen and carbon) that is called "crude oil."

The word "petroleum" means "oil from the earth," or "rock oil." As the earth's geology shifted over millennia, oil and natural gas were trapped in underground pockets. In regions where the rock is porous, oil can be trapped in the rock itself. Man has used oil for approximately 5,500 years to produce heat and power and for many other purposes.[15] The ancient Sumerians, Assyrians, and Babylonians used crude oil and "pitch" (what we call asphalt) from a natural seep at what is now Hit, an Iraqi city on the Euphrates River; oil was also used to build ancient Babylon. In North America, Native Americans used oil to treat illness, to waterproof canoes, and to protect themselves from frostbite. As America grew, petroleum was used to fuel lamps for light. When whale oil became expensive, petroleum oil began to supplant it as a fuel. At the time, most oil was made by distilling coal into a liquid, or was skimmed off of lakes, when petroleum leaked to the surface from underwater seeps.[16]

On August 27, 1859, Edwin L. Drake struck oil in Titusville, Pennsylvania, setting off what has become known as "the Oil Age." Drake pumped oil from underground into wooden barrels. As the oil business grew in the nineteenth century, producers emulated distillers, who transported whiskey in 40-gallon barrels; oilmen copied the idea, adding two gallons to account for spillage. This was significant because it marked the first time purchasers knew exactly how much oil they were acquiring. Although oil is no longer transported in barrels, it is still measured in "barrels" (bbls), or the equivalent of 42 gallons.[17]

Today, oil is the world's most popular fuel, representing 33.1 percent of global energy use.[18] The EIA predicts global demand will jump from 98 million barrels a day in 2020 to 112 million in 2035.[19] Nevertheless, the use of other fuels is also surging, and oil has been losing market share. According to British Petroleum (BP), oil's market share in 2012 was at its lowest point since the company began compiling data in 1965.[20]

To access oil, energy companies drill deep into the earth, then pump the oil from deposits to the surface. It is sent to refineries by pipeline, ship, or barge. Crude oil is considered "sweet" when it contains a small amount of sulfur, and "sour" when it contains a lot of sulfur. Crude is further classified as "light," which flows easily like water, or "heavy," which is thick and viscous like tar.[21]

Oil refineries break down hydrocarbons into various commodities, known as "refined products," such as gasoline, diesel fuel, aviation fuel, heating oil, kerosene, asphalt, lubricants, propane, and the like. Oil can be converted into naphtha, which is the "feedstock," or basis, for high-octane gasoline or lighter fluid. Oil is used to produce many other products, including fertilizers and plastics. As it is processed, oil expands (much like popcorn growing bigger as it pops). A 42-gallon barrel of crude oil generally produces 45 gallons of petroleum products.[22] The vast majority—about 70 percent—of US petroleum consumption is used for transportation.[23]

Gasoline—commonly referred to as "gas" in the United States (and "petrol" in Great Britain), but not to be confused with natural gas—is a fuel made from petroleum. At US refineries, gasoline is the main product produced from crude oil. In 2011, Americans used 367 million gallons of gasoline per day, the equivalent of more than a gallon of gas per day for every citizen.[24] One 42-gallon barrel of refined crude oil will produce about 19 gallons of gasoline. Most gas is used by cars and light trucks, but it also fuels boats, farm equipment, and recreational vehicles.

What Is Natural Gas?

Natural gas is a pure form of fossil fuel composed of methane, or CH_4, a chemical compound made up of one carbon atom and four hydrogen atoms. Natural gas is lighter than air, has no natural odor (we mix it with mercaptan, a chemical with a strong sulfur odor, as a warning of leakage), is often found near petroleum deposits deep underground, and is highly flammable.

The first discoveries of natural gas were made in what is now Iran, 6,000 to 2,000 years BCE (Before the Common Era).

Natural gas seeps there were probably ignited by lightning, and fueled the "eternal fires" of the fire-worshipping ancient Persians.[25]

To find gas deposits today, geologists study seismic surveys—using echoes sent by a vibrating pad under a specially built truck—to identify natural gas deposits deep below ground. At a promising site, a drill rig bores test wells. Once a productive deposit is found, gas is pumped to the surface and is sent to storage tanks by pipeline.

Once stored, natural gas is measured by volume: a cubic foot (cf) of gas is equivalent to the amount of gas that fills a one cubic foot volume, under set conditions of temperature and pressure. A "therm" is equivalent to 100 cf; and "mcf" is equivalent to 1,000 cf. To help compare fuels, energy content is measured in BTUs, or British Thermal Units. One BTU is the amount of heat required to raise one pound of water (about a pint) one degree Fahrenheit at its point of maximum density. One cubic foot of natural gas releases approximately 1,000 BTUs of heat energy (and one barrel of oil equals 6,000 cubic feet of natural gas).[26]

Natural gas requires minimal processing. "Wet" natural gas contains liquid hydrocarbons and nonhydrocarbon gases, such as butane and propane, which are known as byproducts. Once the byproducts are removed, the methane is classified as "dry"—or "consumer grade"—natural gas, and is widely distributed.[27] The production and use of natural gas creates fewer emissions than oil and coal.

Conventional natural gas wells tap into large free-flowing reservoirs of trapped gas that can be accessed by a single, vertical well. In "shale gas" and "tight gas" fields, the gas is trapped in tiny bubbles embedded in dense (or tight) rock formations that are accessed by a combination of vertical and horizontal wells and hydrofracking, as we will see in chapter 2.[28]

In the early days of US energy production, natural gas was considered an unwanted byproduct of oil extraction. (In many parts of the world natural gas is still "flared off," because it requires too much work to collect, transport, and use.) After an enormous gas field was found in the Texas Panhandle in 1918, it was used to manufacture carbon black, which in turn was used to make car tires. Eventually, Americans used natural gas to heat their homes and to fuel power plants.[29] But not until the last decade did it become as important as coal, oil, or nuclear power. Today, natural gas is used for generating electricity (36 percent) and for industrial (28 percent), residential (16 percent), and commercial use (11 percent). The remaining 9 percent is used by energy industry operations, pipelines, and vehicle fuel.[30] Gas is important for manufacturing steel, glass, paper, clothing, and many other goods, and provides raw material for plastics, paints, fertilizer, dyes, medicines, and explosives. It can be converted into ethane, a colorless and odorless gas that is important to the chemical industry. Natural gas provides heat for over half of America's homes and commercial establishments, and it powers stoves, water heaters, clothes driers, and other appliances.

A series of energy-related crises have made natural gas increasingly popular. Coal-fired power plants have lost favor because of their sooty emissions. Deep-water oil exploration faced a setback with the blowout of a BP oil well in the Gulf of Mexico in 2010, which set off the largest oil spill in US history. And an earthquake and tsunami in Japan in 2011 raised questions about the safety of nuclear plants (nuclear plants are also much more expensive to build, operate, and maintain than equivalent natural gas facilities).

Nevertheless, traditional concerns about natural gas included limited supplies and volatile prices—until fracking, that is, which has significantly increased known gas reserves and lowered prices. In July 2008 shale gas cost $13.68 per million BTUs (MBTU) at Henry Hub, a concentration of pipelines in Louisiana that serves as the main pricing point for American natural gas.[31] Thanks to the rapid proliferation of hydrofracked wells, natural gas supplies ballooned at just the moment that the economy—and demand—slacked off, sending gas prices plummeting. Between 2008 and 2012, gas prices fell over 60 percent.[32] After tumbling below $2 per MBTU in 2012, prices doubled to $4.04 in mid-2013 and are expected to rise slightly higher in 2014.[33]

A number of other factors—including ready capital, a trained workforce, greater access to pipelines by third parties, the ability to hedge the risks of gas exploration, and a robust energy market—helped to accelerate gas's acceptance. Utilities saw it as a stable, affordable, relatively clean source of power. In 2008, natural gas amounted to about 20 percent of the nation's energy production; by 2012 it amounted to over 30 percent, a number that is likely to keep growing. The EIA forecasts that domestic natural gas production will grow 44 percent—from 23 trillion cubic feet (tcf) to 33.1 tcf—between 2011 and 2040.[34]

Why Is Natural Gas Called a "Bridge Fuel"?

Natural gas emits less CO_2 than other fossil fuels and requires less processing (or refining, to remove the other elements) than oil, and so it has been promoted as a "bridge fuel"—a cleaner-burning alternative to oil and coal that will ease the transition to renewable energy supplies such as wind, solar, and hydropower.[35]

It will take decades and billions of dollars to scale-up the capacity of renewable energy plants. In the meantime, natural gas advocates—such as the Texas energy billionaire T. Boone Pickens—believe it is a "second best" solution, one that will reduce greenhouse gas emissions in the near term while delivering energy independence from petro states.[36]

Yet some environmentalists deride natural gas as a "bridge to nowhere."[37] Humans have already pumped so much carbon into the air, skeptics say, that switching to natural gas will only cut the global warming effect by 20 percent over 100 years. "There are lots of reasons to like natural gas, but climate change isn't one of them," said climatologist Ken Caldeira and former Microsoft chief technology officer Nathan Myhrvold, who together published a critical study of the gas boom in *Environmental Review Letters*. "It's worthless for [fighting] climate change."[38] Nonetheless, energy executives tout so-called unconventional fuels, many of which are extracted by hydrofracking.

What Are "Unconventional" Fuels?

Unconventional gas refers to shale gas, coal bed methane, subterranean coal gasification, or tight gas.

Unconventional oil is oil acquired by means other than drilling a traditional borehole in the ground. It includes oil shale, oil sands (tar sands), heavy oil, oil synthetically produced from coal or natural gas, natural gas liquids, and lease condensate (a liquid byproduct of natural gas extraction at the well, or "lease," site).[39]

Thanks to horizontal drilling and hydrofracking, shale oil production has grown to provide 29 percent of total crude oil production the United States, and 40 percent of total US natural gas production, according to the US government's Energy Information Administration (EIA).[40]

In a 2013 study of 41 countries, the EIA raised its estimates of global shale oil and shale gas reserves. According to the new report, 10 percent (or some 345 billion barrels) of the world's recoverable crude oil resources, and 32 percent (or 7,299 trillion cubic feet) of global natural gas resources, are found in shale formations.[41] (These estimates remain uncertain until more data become available.) More than half of the shale oil resources outside the United States are concentrated in four countries: Russia, China, Argentina, and Libya. More than half of the shale gas outside of the United States is found in China, Argentina, Algeria, Canada, and Mexico. For now, the United States and Canada are the only nations producing shale oil and gas in commercial quantities.[42]

Unconventional fuels tend to be more difficult to reach, more expensive to produce, and more polluting than conventional fossil fuels. These factors have made them a target of environmentalists. But as conventional supplies become depleted, unconventional fuels will prove increasingly important.

What Is Shale Gas?

Shale gas is natural gas that is trapped in minute deposits embedded in shale rock formations (much the way small air pockets form in a loaf of bread as it bakes). Shales are fine-grained sedimentary rocks. There are different kinds of shale, but the most common is "black shale," which formed in deep water and without any form of oxygen present. Shale gas is usually accessed through a combination of vertical and horizontal drilling, and hydrofracking.[43]

About 33,000 natural gas wells are drilled in the United States annually, and today nine out of ten of them use fracking.[44]

The recent boom in shale gas has been phenomenal. As recently as 2000, shale gas provided only 1 percent of the natural gas produced in America. By 2010 it was over 20 percent, and by 2035 the EIA estimates that 46 percent of US natural gas will come from shale gas.[45]

Today, "Technically recoverable unconventional gas"—which refers to shale gas, tight sands, and coal bed methane—account for 60 percent of America's onshore reserves. According to the EIA, these resources could supply the nation for the next 90 years. (Other estimates predict the supply could last even longer.)[46]

What Is Tight Gas?

Tight gas is natural gas that is trapped in areas of low-porosity silt and sand. Tight gas has less than 10 percent porosity and less than 0.1 millidarcy of permeability.

Porosity is the proportion of void space to the total volume of rock. While beach sand has a porosity of about 50 percent, tight gas is trapped in pores up to 20,000 times narrower than a human hair.

Permeability is the ability of fluid to move through pores. A person can blow air through rock with 1,000 millidarcies permeability.[47]

What Is Sour Gas?

In certain regions, like the Rocky Mountains, natural gas is mixed with elevated levels of sulfur, which creates the corrosive gas H_2S, or hydrogen sulfide. Known as sour gas, H_2S requires extra processing to purify it.

What Is Shale Oil?

These are heavy, viscous crude deposits that cannot be produced and refined by conventional methods. Heavy crude oils usually contain high concentrations of sulfur and a number of metals, particularly nickel and vanadium. These properties make them difficult to pump out of the ground or through a pipeline and interfere with refining. These properties also create serious environmental challenges.[48] The best-known reserve of shale oils is Venezuela's Orinoco heavy oil belt, which contains an estimated 1.2 trillion barrels.

What Is Oil Shale?

This consists of "tight" (dense) formations of sedimentary rock that are not permeable enough to allow the pumping of the oil trapped inside. This oil is in the form of kerogen, a solid mixture of organic compounds. Extracting oil from kerogen is difficult and expensive, and to date oil shale deposits have not been widely developed. But there is an estimated 2.8 to 3.3 trillion barrels of oil in oil shale deposits worldwide, 62 percent of which are in the United States. Aside from hydrofracking, technology for extracting such oil includes igniting the shale underground, which allows light oil to migrate out of the kerogen to pumping stations.[49]

What Are Tar Sands?

Also known as "oil sands," these are deposits of heavy oil so viscous they do not flow. Tar sands are extracted by injecting hot steam into deposits, which heats the sands and makes the tar more liquid so that it can be pumped out. The process requires large amounts of water and energy, is environmentally destructive, and only makes economic sense when oil prices are high. The world's largest tar sands deposits are in the Athabasca region of Alberta, Canada. If approved, the controversial Keystone XL Pipeline would transport oil from the Athabascan fields across the United States to terminals on the Gulf Coast.[50]

What Is Coal Bed Methane?

This is a natural gas extracted from coal beds. It is usually produced by drilling a borehole into a coal seam, reducing the pressure of water flowing through the seam, and allowing the gas held in the seam to flow up the borehole to the surface.

What Is Coal Gasification?

This is a process used to produce "syngas" (synthesis gas), a fuel mixture of hydrogen, carbon monoxide, and carbon dioxide. Originally used in the eighteenth century to create heat and light, modern gasification plants supply fuel for a wide range of uses. One advantage of gasification is that carbon dioxide (the leading greenhouse gas) and other pollutants can be removed from the gas before it is burned. Yet gasification remains more expensive than coal itself, and while it is used for industrial purposes it has not been widely used as a fuel source yet. Research is ongoing to develop gasification plants that will capture and store carbon dioxide, a process that will raise capital costs but reduce climate-changing emissions.[51]

What Is Synfuel?

Synthetic fuels, or "synfuels," are liquid fuels converted from coal, natural gas, oil shale, or biomass (plant-based fuels, such as wood, sugarcane, or algae). Conversion of coal to synthetic fuel was pioneered in Germany a century ago, and the Nazis greatly expanded the use of synthetic conversion to produce aviation fuel and oil during World War II. There are several methods, primarily the Bergius process, developed in 1913, in which coal is liquefied by mixing it with hydrogen gas and heating it (hydrogenation); the coal is mixed with oil; and catalyst is added to the mixture.

2

WHAT

So What Is Hydrofracking, and Why Has It Become So Central to the Energy Landscape?

As mentioned in the introduction, hydrofracking is defined in different ways by different people. To those in the energy industry, it refers purely to the process of injecting fluid—which consists of water and between 3 and 12 chemicals, including hydrochloric acid, sodium chloride, methanol, and isopropyic ethanol, that serve various functions—underground at high pressure to crack open shale rock, release natural gas or oil trapped there, and allow the hydrocarbons to flow to the surface.[1] Energy specialists will point out that hydrofracking and drilling are not the same thing. First comes drilling of the wells, then comes the hydrofracking.

But such distinctions are lost on the public. To most people, hydrofracking and drilling are simply two parts of one process. In this broader definition, "fracking" is a shorthand way to describe all of the steps used to prepare a well, drill it vertically and horizontally, inject the fluid, recover hydrocarbons, and remediate the waste. For the sake of clarity and brevity, I have used the latter, broader definition in this book.

Fracking has emerged as a central issue in part because of the significant changes we are seeing in today's energy landscape. According to the EIA, total energy consumption in 2008 by the global population of seven billion was 493 quadrillion BTUs—or 493,000,000,000,000,000 BTUs.[2] As the global population expands to nine billion by 2040, the demand for affordable, reliable, plentiful energy supplies will increase exponentially. Daniel Yergin, a leading energy consultant and Pulitzer Prize–winning author of books such as *The Quest: Energy, Security, and the Remaking of the Modern World*, predicts that global energy demand will increase 35 to 40 percent in the next two decades alone. "Much of the infrastructure that will be needed in 2030 to meet the energy needs of a growing world economy is still to be built," he has said.[3]

The United States has long held the title for the world's largest economy, and has been as well its foremost energy user. With less than 5 percent of the world's population, the United States consumes 20 to 25 percent of the world's supply of fossil fuels.[4] But in 2010 China surpassed the United States to become the globe's biggest energy consumer: the United States uses 19 percent of the world's energy, and China accounts for 20.3 percent of global energy use.[5] By 2016, the International Monetary Fund (IMF) forecasts, China will eclipse the United States as the world's leading economy.[6]

As concern about rising global temperatures, high oil prices, the vulnerability of nuclear plants, and friction over energy supplies grows, there will be calls for greater use of "renewables"—sunlight, wind, moving water, and geothermal heat. A shift to new technologies and fuels is already under way, and defining what resource economists call "a new age in the history of energy."

The American public, for one, seems to favor a cautious approach. In the US presidential election of 2012, Republicans and Democrats sparred over energy policy, among many other issues. But political theater aside, Obama and Governor Mitt Romney agreed on most issues: both said they wanted to lower dependence on oil from the Middle East and to encourage more oil drilling in Alaska, Texas, North Dakota, and the Gulf of Mexico. Both said they supported biofuels and nuclear power. And each candidate asserted that he would promote hydrofracking better than his competitor.

What Is Hydrofracking Used For?

While the public generally associates hydraulic fracturing with natural gas, the process is also used for many other purposes, including

- Extracting oil and "liquid" natural gases (LNGs), such as propane, butane, hexane, and the like
- Preventing mining cave-ins
- Stimulating or gauging groundwater flow
- Accelerating the flow of water from drinking supplies
- Injecting wastewater (including from hydrofracking) into deep rock formations for permanent storage
- Disposing of oil or gas waste, and remediating Superfund sites (i.e., places with the highest levels of pollution in the country)
- Extracting geothermal heat from underground to produce electricity
- Increasing injection rates for geologic sequestration of CO_2[7]

What Is the History of Hydrofracking?

Originally, vertical wells tapped oil- or gas-bearing limestone or sandstone formations, which are relatively porous. Natural gas was first extracted from shale rock in Fredonia, New York, in the 1820s by means of conventional drilling. The first fracturing of shallow, hard-rock wells occurred in the 1860s, when prospectors in Pennsylvania, New York, Kentucky, and West Virginia used nitroglycerin (NG) to search for oil in shale formations. Though NG could explode and its use was illegal, the technique was extremely successful. Later, the same methods were employed in water and gas wells. But it was Floyd Farris, of Stanolind Oil and Gas Corporation (an exploration and production business started by Indiana Standard in 1931, which later became Amoco, now part of BP), who conducted experiments in the 1940s and saw the potential of hydraulic fracturing to enhance production from oil and gas wells. The first attempt to "hydrafrac" a well took place in the Hugoton gas field in Grant County, Kansas, in 1947: Stanolind injected 1,000 gallons of gelled gasoline and sand taken from the Arkansas River into a gas-bearing limestone formation 2,400 feet deep. The experiment was not a huge success, but it led to further experiments by Stanolind. A patent on the process was issued in 1949, and an exclusive license for its use was granted to the Halliburton Oil Well Cementing Company (Howco) of Texas. Howco performed the first two commercial hydraulic (water-based) fracturing treatments, in Stephens County, Oklahoma, and Archer County, Texas. The process was used on 332 wells in 1949, and production increased a surprising 75 percent. At first, crude oil, gasoline, and kerosene were used as fracturing fluids. But in 1953, they were replaced by water.[8]

Chemists developed gelling agents to fine-tune the viscosity of fracking fluids, which helped the wells perform more efficiently. A series of advances since then have made the chemistry and engineering of hydrofracking highly sophisticated.

The high porosity and low permeability of shale makes it a difficult medium to work in. It took years of research by government and industry to develop modern hydrofracking techniques. In the 1970s the federal government initiated the Eastern Gas Shales Project and dozens of pilot hydrofracking projects, and supported public-private research. These efforts were spurred by the energy crisis of 1973, when the Arab members of OPEC, the Organization of Petroleum Exporting Countries, imposed an oil embargo to punish the United States for its support of Israel during the Yom Kippur War.[9] It was also spurred by the decline in conventional natural gas production in the United States over the course of that decade. In response to the crisis, the Ford and Carter administrations prioritized the search for new energy supplies. Industry and federal researchers began to focus on techniques to access "unconventional" resources, such as those listed above—coal bed methane, "tight sands" natural gas, and shale gas.

The National Labs—Sandia and Los Alamos in New Mexico, and Lawrence Livermore in California—provided computer modeling, monitoring, and evaluation to demonstration projects. In 1979, the public-private efforts to drive shale gas and coal bed methane to market were formalized in the Department of Energy's Commercialization Plan for Recovery of Natural Gas from Unconventional Sources. New, three-dimensional microseismic imaging, originally developed by the Sandia Lab for coal mines, was used to locate fractures in shale and "see" unevenly distributed gas formations deep underground. In the meantime, researchers found that diamond-studded drill bits were far more effective at boring through tough shale than conventional drill bits.[10]

In 1980, a year after the second energy crisis in 1979 caused by the Iranian Revolution, Congress passed the Windfall Profits Tax Act, which created the Section 29 production tax credit for unconventional gas. By providing an incentive of $0.50 per thousand cubic feet of natural gas produced from unconventional resources, the tax credit spurred the growth of shale hydrofracking.[11]

Most of the early research and demonstration work was conducted in the Devonian and Marcellus Shales, both of which are located on the East Coast. But the key breakthrough happened in the Barnett Shale, in Texas, where a determined tinkerer solved the vexing problem of how to coax gas to flow out of a hydrofracked well smoothly and quickly.

Who Was the Determined Tinkerer?

George Phydias Mitchell was the son of a Greek goatherder (Savvas Paraskevopolous, who changed his name to placate an American employer) who became a Galveston-based oil and gas producer. George Mitchell and his brother drilled 10,000 wells around Fort Worth, and became successful at working oil fields others had given up on. In the 1980s Mitchell used government mapping tools and other research to learn how to hydrofrack the Barnett Shale formation. But this shale formation is complex and drillers were stymied by slow, clogged, inefficient wells. Mitchell spent 17 years and $6 million to crack the problem, though many told him he was just wasting his time. It was, says *The Economist*, "surely the best development money in the history of gas."[12]

Having successfully demonstrated multifracture horizontal well drilling, Mitchell Energy engineers had to develop the optimal combination of inputs—water, sand, proppants, chemical lubricants, and so on—to achieve maximum gas recovery at the lowest cost. In 1997, Mitchell Energy developed "slickwater fracturing," in which chemicals are added to the water pumped into wells to increase the fluid flow. The chemicals, some of which are discussed in the next chapter, had particular functions—friction reducers, biocides, surfactants, and scale inhibitors—to prevent a well from clogging, keep proppants suspended, and accelerate gas pumping. The invention of the slickwater process was the breakthrough that made shale gas economical, and in 1998 Mitchell tapped huge shale gas reserves. The *Oil and Gas Journal* notes that as a result of Mitchell's refinements of fracking techniques, a well that produced 70 barrels a day using conventional (vertical) drilling can now produce 700 barrels a day. And a hydrofracture job that once cost $250,000 to $300,000 now cost about $100,000. Passing this milestone, shale gas became a commercially viable resource.[13]

Mitchell's technology is now used nationwide, both for gas and for shale oil, which can be extracted from shale beds in the same way gas is. Some wells also produce valuable liquid natural gases (LNG), such as butane and propane. "We can frack safely if we frack sensibly," Mitchell wrote with Michael Bloomberg in the *Wall Street Journal*.

Drilling and hydrofracking techniques continue to evolve, and since the 1940s, roughly 1.2 million wells have employed them in the United States.[14] By 2009, there were nearly 500,000 active natural gas wells in the United States, double the number in 1990, and the drilling industry reports that about 90 percent of them used hydrofracturing (others say the figure could be as high as 95 or even 99 percent.[15])

The Barnett Shale produces over 6 percent of all domestic natural gas.[16] In 2001, just before the Section 29 production tax credit expired, George Mitchell sold his company to Devon Energy for $3.5 billion.[17] (He died at age 94, in July 2013.)

3

HOW

How Do We Hydrofracture a Well?

Let's begin with the equipment. Roughnecks rely on tall metal drill rigs (such as the rig depicted on the cover of this book) that rise up to four stories tall, which lower diamond-tipped drill bits and sections of steel pipe into the borehole.

Other equipment includes a slurry blender, high-volume fracturing pumps, a monitoring unit, fracturing tanks, proppant storage and handling units, high-pressure treating iron, a chemical additive unit, low-pressure flexible hoses, and gauges and meters to assess flow rate, fluid density, and treating pressure.[1]

Once a drill pad has been built and the equipment is in place, drillers use a series of choreographed steps to hydrofrack an oil or natural gas well.[2] The first step is to drill a vertical borehole into a layer of shale, which typically lies a kilometer (3,280 feet, or 0.62 of a mile) or more beneath the surface.[3] This depth can vary widely, depending on the location, geology, stage of drilling, and so on. In the Marcellus Shale, for instance, natural gas wells range from 5,000 to 9,000 feet deep. (By comparison, most residential water wells lie 200 to 500 feet deep.)[4]

The second step is to reach that depth and turn the drill bit horizontally and continue drilling, extending the lateral borehole up to a mile or two long.

The third step is to line the vertical and horizontal borehole with steel casing, to contain the gas and (so it is hoped) to protect groundwater from pollution, and cement it in place.

The fourth step is to use explosives to perforate holes into the horizontal section of the well casing. This is done by detonating a small package of ball-bearing-like shrapnel with explosives; the shrapnel pierces the pipe with small holes.

The fifth step "completes" the process by using powerful pumps to inject "slickwater" fluids—a slurry (a semiliquid mixture of soluble and insoluble matter) consisting of water, sand, and chemicals—into the wellbore at extremely high pressure (over 9,000 pounds per square inch).[5] When the pressurized fluid flows through the perforations at the end of the wellbore, it fractures the shale rock in which gas is trapped in tiny bubbles. As the fluids break open the rock, sand and other "proppants"—materials that hold the fractures open—allow gas to flow out; the chemicals help the natural gas escape the rock and flow up the borehole to the surface.

The sixth step is the "flowback" phase, in which pump pressure is released, allowing much of the fluid in the well to return to the surface.

The seventh step involves cleaning up the borehole and allowing the well to start producing gas. This can take several days.

In preparing a well for production, as many as 25 fracture stages may be used, each of which uses more than 400,000 gallons of water—for a total, in some cases, of over 10 million gallons of water—before a well is fully operational.[6]

The eighth step is to remediate the wastewater a well produces, which I discuss below.

How Do Horizontal Wells Differ from Vertical Wells?

To answer this, we need to return to the first three steps listed above—the drilling. To hydrofrack a shale rock formation, boreholes are generally drilled straight down some 5,000 to

20,000 feet deep. The average depth is 7,500 feet deep, which is one-and-a-half miles below the surface—equivalent to six Empire State Buildings or more than 25 football fields stacked up end to end.[7] (The depth is key to environmental safety, as the industry points out that these wells extend beneath the water table.)

Since the 1990s, hydrofracking has been combined with "directional" drilling, in which a vertical well is drilled thousands of feet deep, whereupon the borehole doglegs and continues horizontally through layers of gas- or oil-bearing shale. In places like the Bakken Formation, in North Dakota, the lateral leg can extend for one or two miles.[8]

A horizontal well therefore begins as a standard vertical well. A rig will drill thousands of feet down, until the drill bit is perhaps a few hundred feet above the target rock formation. At this point the pipe is pulled from the well, and a hydraulic motor is attached between the bit and the drill pipe, and lowered back down the well. The hydraulic motor is powered by

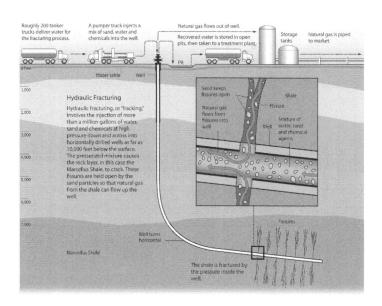

Al Granberg/ProPublica

forcing mud (or slurry) down the drill pipe. The mud rotates the drill bit without rotating the entire length of pipe between the bit and the rig, allowing the bit to change direction, from drilling vertically to horizontally.[9]

Operators on the surface carefully monitor instruments lowered into the well to monitor the azimuth—the measurement of angle in a spherical context, like the earth—and orientation of the drilling, and steer the drill bit underground.[10] Three-dimensional seismic imaging technology has made it easier for drillers to identify "sweet spots," places where gas has collected in large reservoirs. Once the borehole has been steered into the shale, straight-ahead drilling resumes, moving horizontally through thousands of feet of shale rock.

The combination of directional drilling and hydraulic fracturing works well in places like the Marcellus Shale and Bakken Formation, which were not productive using traditional drilling techniques. From the industry's perspective, the advantages of directional drilling are many.

First, horizontal drilling allows companies to hit oil or gas reservoirs that are difficult to access, and to stimulate them in ways that simple vertical wells cannot. Some shale reservoirs are located in places where drilling is difficult or not allowed— such as under a school or city. But if drill pads are located on the edge of a city, such as those that ring the Dallas-Fort Worth Airport (DFW), in Texas, and their wells are drilled at an angle that intersects the reservoir, it is feasible—and legal—to tap into it.

Second, horizontal drilling permits the drilling of multiple wells—often six at a time—from a single pad, reducing the cost and environmental impact of energy projects. In 2010, the University of Texas at Arlington made headlines when researchers there ran 22 wells from a single pad to pull natural gas from a 1,100-acre shale formation beneath the campus. (The pad is on state-owned land, and isn't bound by local municipal rules governing urban gas drilling. Carrizo Oil & Gas operates the wells on university property, which have produced some

110 billion cubic feet of gas).[11] But such schemes can be controversial. In 2013, students, citizens, and environmentalists protested the fracking of gas beneath a pristine, biologically rich, 8,000-acre tract of forest land owned by the University of Tennessee in Knoxville.[12]

Additionally, while horizontal drilling is technically complex, time-consuming, and expensive—a directionally drilled, hydrofracked shale well can cost two or three times as much per foot as a standard vertical well—the wells tend to be more productive than vertical wells, and the extra cost is made up for by bigger gas or oil yields.[13] The reason for this is that the size of a well's "sweet spot" or "pay zone" becomes much larger. A vertical well drilled through a 50-foot thick section of shale will produce a pay zone that is 50 feet wide. If the well turns sideways halfway through the vertical section and runs horizontally for 5,000 feet, then the sweet spot full of gas bubbles will be 5,000 feet long, which greatly increases its productivity.

Horizontal wells can also be safer. In the case of a well blowout—which occurs when pressure-control systems fail, and natural gas is released in an uncontrolled way (blowouts can be explosive and spew fracking fluid into the air and onto the ground)—directional drilling can provide a relief well, which intersects it, allowing rig workers to relieve pressure and control it or seal it.[14]

Finally, horizontal drilling also has uses other than for hydrofracking. It can be used to install utility lines that cross beneath a city, road, or river, for example.

It is important to note that vertical and horizontal wells are subject to different regulations. In the case of a vertical well, gas is taken from beneath a single piece of surface property. Most states and territories have fairly straightforward mineral rights rules that govern the ownership of gas or oil produced from vertical wells. In the case of horizontal wells, a single borehole can cross beneath numerous pieces of property, perhaps owned by people with differing agendas. In general, the royalties from horizontally drilled wells are established before drilling begins through a combination of government rules and often complex private royalty-sharing agreements. But different states have different laws.

Like most Western states in the United States, for instance, Colorado makes a legal distinction between surface property rights and subsurface mineral rights.[15] This distinction is based on Spanish legal precedent (East Coast law, on the other hand, tends to be based on the British precedent.) In Colorado—or North Dakota or Texas—a real-estate transaction may or may not include mineral rights.[16] In those states, it is common for the minerals beneath a property to be sold, leased, or retained by previous owners. But there is no law requiring owners to tell buyers about the disposition of mineral rights; there have therefore been many cases in which a buyer did not realize the previous owners had leased the mineral rights to energy companies. In the early days of mineral exploration, this kind of situation reliably led to disputes or even violence.

To obtain the right to drill under private property, an energy firm leases the mineral rights from the federal Bureau of Land Management (BLM).[17] If the lease is granted, the company can drill without the surface property owner's permission. The Colorado Oil and Gas Conservation Commission (COGCC) is responsible for promoting energy development, and tends to favor drillers because they generate revenue for state coffers. Between 2008 and 2019, Colorado estimates it will earn some $2.7 billion from energy extraction. The COGCC also fines drillers for violations of its rules protecting human health and safety: between 1994 to 2000, the Commission collected $1 million in fines from 110 violations.[18]

But citizens groups worry that fracking will damage homes, harm wildlife, and affect outdoor recreation, and have petitioned the COGCC to stop, or rescind, drilling permits. This type of "resource war" will become increasingly common as the population grows, the climate changes, and the demand for energy pushes fracking into new regions.

What Are Hydrofracking Fluids?

Having drilled a directional well, we can turn to what gets injected into them. I've mentioned some of the chemicals, as well as the "proppant" that is carried into the shale formation to help release the oil or gas trapped there.

First, acid scours the borehole and the fractures in the rock. Then fluid is injected into the borehole, with the pressure greater than the fracture gradient of the rock. The fluid includes water-soluble gelling agents, such as guar gum, which increase viscosity and help deliver proppant to the bottom of the well.[19] As the fracturing proceeds, viscosity-reducing agents—such as oxidizers and enzyme breakers—are added to the fluid to deactivate the gelling agents.

The components of fluid vary, depending on the specifics of the site, but are typically 99 percent water and sand and 0.5 to 1 percent chemicals.[20]

Drillers usually begin by pumping hydrochloric acid or acetic acid to dissolve rubble, clean the borehole, and reduce pressure on the surface. Then the proppant is added to the fracking fluid and sent down the well. Proppants are small particulates—typically grains of silica sand, resin-coated sand, or man-made ceramic balls—that are used to prop open the fractures in shale rock and keep them open, even after fracking is completed. In essence, proppants make shale artificially permeable.

Drillers use two methods to deliver proppants to the bottom of a well: high rate and high viscosity. High-rate, or slick-water, fracturing causes small spread-out microfractures. High-viscosity fracturing tends to cause large dominant fractures in the rock.

The chemicals in the fluid are tailored to the specific geology of the site, so as to protect the borehole and improve the flow of oil or gas to the surface.[21] They include fresh water, salt water, foams, friction reducers, and gelling chemicals. The friction reducer is usually a polymer, which reduces pressure loss caused by friction, allowing pumps on the surface to work at a higher rate without adding greater pressure.

A number of chemicals are used to increase the viscosity of the fracturing fluid, which carries proppant into the formation. But to stop the fluid from pulling proppants out of the formation, a chemical known as a "breaker"—usually an oxidizer or enzyme—is pumped with gel or cross-linked fluids to reduce the viscosity. For a complete list of the chemicals used in hydrofracking, see the appendix at the back of this book.

At the end of fracking, a well is commonly flushed with water, sometimes blended with a friction reducer, under pressure. Some of this wastewater is recovered after fracking and must be carefully disposed of (more on this below).

Water remains by far the largest component of fracking fluid. The initial drilling operation alone may require some 6,000 to 600,000 gallons of fluids.[22] According to the EPA, the total volume of water used to hydrofrack a well ranges between 65,000 gallons, such as for shallow coal bed methane production, to 13 million gallons for deep-shale gas production.[23] Most wells use between 1.2 and 4 million gallons of fluid, with large projects using up to 5 million gallons (equivalent to the amount of water used by approximately 50,000 people during the course of one day).

Once the hydrofracking fluid has been injected and pressure from the pumps is released, fracturing fluid known as "flowback" surges back up through the borehole to the surface. "Produced water" is fluid that returns to the surface once the well has started producing natural gas or oil. Collectively, flowback and produced water are known as hydrofracking "wastewater," which is suffused with salts, chemicals, debris scoured from the wellbore, and even naturally occurring radioactive elements.

Because it is contaminated, the question of how to capture, store, and treat millions of gallons of wastewater is a knotty one, and the industry has resorted to several different strategies. In western and southern states, wastewater is often injected back underground into storage wells; when injected into geologi-cally active zones, this has occasionally set off minor tremors.[24] Some communities allow wastewater to be spread on roads or fields, for dust suppression or de-icing. Environmentalists worry that the toxins will wash into freshwater and food sup-plies. In certain cases, wastewater is recycled for use in other wells. In eastern states like Pennsylvania, much of the waste-water is shipped by pipeline or truck to public sewage treat-ment plants; but those plants are generally not equipped to process the chemicals or the naturally occurring radioactive material that is sometimes dredged up. The improvement of wastewater recycling and disposal methods is a major focus of the industry. (I discuss the concerns about wastewater and earthquakes at greater length in chapter 6.)

4

WHERE

What Are Shale Plays, and Where Are the Major Shale Plays in the United States?

As mentioned, the purpose of hydraulic fracturing is to access natural gas and oil trapped in shale formations, also known as "plays." Shale plays are found across the United States and around the world, including in large deposits in South America and China, where hydrofracking is only in the experimental stage at the moment.

The United States' most famous and productive shale play thus far is the Barnett Shale, which spreads for 5,000 square miles around Fort Worth, Texas, and provides 6 percent of the nation's natural gas.[1] (The Barnett is where Mitchell Energy perfected slickwater fracking.) Since 2005, thousands of people who live on top of the Barnett have leased their land to drillers and become wealthy. By 2008, landowners in the southern counties earned bonuses of $200 to $28,000 per acre, with royalties ranging from 18 percent to 25 percent (one lease in Johnson County was permitted for 19 wells).[2] But the Marcellus Shale, which spreads beneath five states in the Northeast, is thought to be from two to four times the size of the Barnett, and home to the world's second-largest gas deposit after the South Pars field, beneath Qatar and Iran.[3] Development of the Marcellus Shale has only just begun. Range Resources, based in Fort

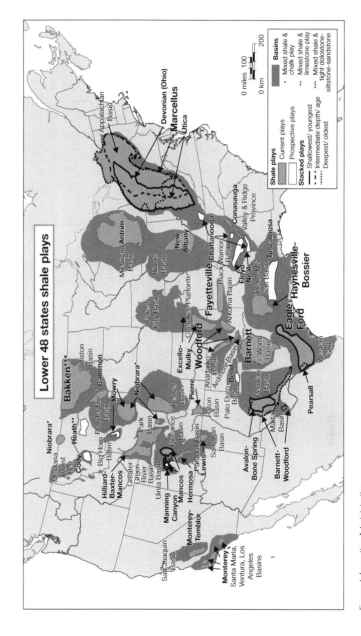

Lower 48 states shale plays

Energy Information Administration

Worth, Texas, has worked both plays: in 2011, it sold off all of its holdings in the Barnett Shale and invested the $900 million of the proceeds in the Marcellus. "Barnett is a good play, but the quality of the rock and economics don't really complete with the Marcellus," said Range CEO Jeff Ventura.[4]

While there are numerous plays spread out across the country, the top few represent the majority of land and have attracted the largest drilling and development investment, which totaled more than $54 billion in 2012.[5] These plays are focused on oil or liquid gas, with the exception of the Marcellus, where the majority of wells are dry gas completions. (As mentioned in the fossil fuel primer in chapter 1, "dry" gas is pure methane.)

The major plays in the United States include the following:[6]

Anadarko-Woodford: Located in west-central Oklahoma, this play runs through the Anadarko Basin, at some 11,500 to 14,500 feet deep. It contains crude oil and liquid gas, and the average cost of drilling and completing a well there was $8.5 million in 2012.

Bakken: Covering some 200,000 square miles beneath parts of Montana, North Dakota, Saskatchewan, and Manitoba, this organic-rich, low-permeability formation contains the largest known oil reservoir in the continental United States: an estimated 3.6 billion barrels of oil, 1.85 trillion cubic feet of natural gas, and 148 million barrels of natural gas liquids, according to the US Geological Survey. The play is relatively thin and runs from a depth of 3,000 feet to 11,000 feet deep. For years it was considered uneconomical to drill there, but with the advent of modern fracking this resource trove set off the legendary "Bakken boom," which enriched North Dakotans and brought people from across the country to one of the biggest oil discoveries of recent years.

Barnett: Situated directly beneath the Dallas–Fort Worth metroplex, the Barnett holds some 360 trillion cubic feet of natural gas, and represents one of the biggest shale gas plays in the country. The Barnett is known as a "tight" gas deposit, where the rock is very hard, permeability is low, and the extraction of natural gas is difficult. The combination of horizontal drilling and hydrofracking opened up the Barnett in the late 1990s. Today, with 13,000 wellbores reaching maturity there, companies like Halliburton—which has drilled over 10 million feet of horizontal wells, and fractured and refractured wells in the Barnett—are exploring the region's nooks, crannies, and edges for gas reserves and untapped oil deposits.

Eagle Ford: stretching some 400 miles long, this play underlies southwest and central Texas. The shale here dates to the Upper Cretaceous period, and is brittle (which is good for fracturing). The shale formations occur as a wide sheet some 40 to 400 feet thick, at depths of 4,000 feet to 14,500 feet, and contain oil and some 150 trillion cubic feet of dry and wet natural gas.

Fayetteville: Located in the Arkoma Basin in Arkansas, this is the nation's second-most-productive shale play (after Haynesville, below), is among the United States' 10 largest energy fields, and holds some 20 trillion cubic feet of gas. The Fayetteville covers some 4,000 square miles, ranges in thickness from 60 to 575 feet, and runs from 1,450 feet to 6,700 feet deep.

Granite Wash: A group of plays in the tight sand beneath north Texas and south Oklahoma, where layers of minerals were deposited by ancient streams and washouts. Here a number of oil and gas formations are 1,500 to 3,000 feet thick, and are stacked at 11,000 to 15,000 feet deep.

Haynesville: With 5.5 billion cubic feet of natural gas recovered from the play every day, Haynesville contains an estimated 250 trillion cubic feet of gas, and could surpass the Barnett Shale to become the most productive shale play in the United States. Rich in organic fissile black shale of the Upper Jurassic age, it is located on what is called the Sabine Uplift, which separates the salt basins of east Texas and north Louisiana. The play covers 9,000 square miles, and, running at depths of 10,000 to 14,500 feet, it is far deeper than most plays. Average wells are at some 12,000 feet deep. As a result, temperatures and pressures here are intense: the temperature at the bottom of the well averages 300 degrees Fahrenheit, and pressures at the wellhead exceed 10,000 pounds per square inch.

Marcellus: As noted earlier, this play is located in the Northeast, the biggest gas-consuming region of the country, and was first developed in 2004. For now, most of the Marcellus gas wells are in Pennsylvania, which has aggressively promoted hydrofracking. The wells there average about 6,200 feet deep and cost roughly $5 million each. By contrast, neighboring states have been more cautious to exploit the Marcellus. New Jersey and New York have temporarily banned hydrofracking, pending the results of lengthy environmental and health reviews. The Marcellus is a natural gas shale stacked on top of another major play, the Utica Shale, which contains oil (see below).

Monterey Shale: Running southeast of San Francisco, this formation covers about 1,750 square miles, from southern to central California. The federal EIA estimates that it holds more than 15.4 billion barrels of oil in shale deposits that are estimated to be 1,900 feet thick and lie at an average depth of 11,000 feet. If the estimates prove accurate, that would mean the Monterey contains four times as much tight oil as the Bakken Formation, and approximately 64 percent of the nation's shale oil reserves. If exploited, the Monterey's tight oil could yield 2.8 million jobs and $24.6 billion in state and local taxes, boosting California's economy by 14 percent by 2020, according to a recent study from the University of Southern California. But with complex geology that is prone to earthquakes, disappointing early production numbers for wells, a lack of water, and strong environmental opposition, widespread hydrofracking in California is not a given. (This region has long been commercially exploited by conventional oil drillers, and is the nation's third-highest oil producer, after Texas and North Dakota.)

Niobrara: This emerging oil and gas play lies near the Rocky Mountains beneath northeastern Colorado, Wyoming, Nebraska, and Kansas. Made up of Cretaceous rock, the Niobrara deposit measures between 150 to 1,500 feet thick and lies about 6,200 feet deep. Though exploitation is still in its early stages, the Niobrara's brittle, calcareous chalk is well suited for hydrofracking, and it is considered a potentially huge play. Most production is focused on the northeastern corner of Colorado, in the Denver-Julesburg basin.

Permian Basin: Extending 300 miles long and 250 miles wide and located under New Mexico and west Texas, the Permian has been drilled for oil with conventional rigs since 1925. The formation is filled with Paleozoic sediments, and has one of the world's thickest rock deposits from the Permian geologic period (dating to some 290 million years ago). Today this play is undergoing a renaissance thanks to hydrofracking. It has been estimated that the Permian Basin is rich, with a single formation—the Spraberry—holding 500 million barrels of unconventional oil and five trillion cubic feet of natural gas.

Utica Shale: Lying 3,000 to 7,000 feet *beneath* the Marcellus Shale, the Utica formation is one of the largest natural gas fields in the world. Stretching from Quebec and Ontario in Canada down through New York, Pennsylvania, Ohio, and West Virginia, it also underlies parts of Kentucky, Maryland, Tennessee, and Virginia. The Utica play is estimated to hold some 1.3 to 5.5 billion barrels of oil and about 3.8 to 15.7 trillion cubic feet of natural gas. The Utica Shale is organic-rich calcareous black shale dating to the Late Ordovician period (460 million years ago). It has a high carbonate content and a low clay mineral content, making the Utica rock more brittle than the Marcellus Shale. This requires different hydrofracking strategies and makes Utica wells more expensive to develop than those in the shallower Marcellus formation. Experts consider the Utica "a resource of the distant future."

Woodford: Lying beneath most of Oklahoma, the Woodford Shale has complex mineralogy and geology, which makes drilling a challenge. This play has produced oil and gas since the 1930s, and the first horizontal wells were drilled there in 2004. Today, some 2,000 wells are in production there, including over 1,500 horizontal wells.

Where Is Hydrofracking Restricted?

In spite of the riches buried in the shale plays listed above, hydrofracking can be geologically, technically, and economically challenging. But an even greater obstacle might be citizen and environmental opposition to the controversial process. While advances in drilling techniques and the types of chemicals used have dramatically changed the scenarios for fracking, lawsuits have challenged the Bureau of Land Management's leasing of federal lands to energy companies.[7] Concerns about hydrofracking's impact on global warming, endangered species, wildlife habitat, farm animals, and food supplies have spurred environmental groups to action. And citizens' fears about the impact of industrial processes, truck traffic, and the social impact on rural communities have led several states, and individual towns, to seek bans on fracking. I will discuss this at greater length in chapter 6, but in the context of where hydrofracking is allowed, it is instructive to see where it is not.

In June 2011, New Jersey became the first state whose legislature passed a moratorium on hydrofracking, in a 33 to 1 vote.[8] In January 2012, Ohio lawmakers placed a temporary ban on hydrofracking after a panel of experts argued that the process of storing used fracking fluid, or "flowback," was to blame for creating massive sinkholes and an outbreak of earthquakes—including a 4.0-magnitude temblor that hit on New Year's Eve 2012.[9] In May 2012, Vermont became the first state to ban hydraulic fracturing outright. "This is a big deal," announced Democratic governor Peter Shumlin. "This bill will ensure that we do not inject chemicals into groundwater in a desperate pursuit for energy."[10]

In Colorado, the city of Longmont (population 88,000) banned hydrofracking in 2012, and was promptly sued by Governor John Hickenlooper—who believes that hydrofracking fluid is so harmless that he reportedly drank a vial of it in public. "The bottom line is, someone paid money to buy mineral rights under that land. You can't harvest the mineral rights without doing hydraulic fracturing, which I think we've demonstrated again and again can be done safely," Hickenlooper countered.[11] In defiance of the governor, the city of Fort Collins (population 147,000) banned hyrdofracking in March 2013. Both the governor and the Colorado Oil and Gas Association have threatened to sue the city.[12]

Is Hydrofracking Taking Place outside of the United States?

Significant shale plays have been identified around the world, from Australia to India to Russia, but for now commercial hydrofracking is only taking place in the United States and, to a lesser extent, Canada. There are several reasons for this. First, the process was developed in America, which is now years ahead of every other nation in research, development, and use of fracking techniques. Second, favorable conditions in the United States have allowed fracking to spread quickly. The Council on Foreign Relations has identified seven critical "enablers" of the American shale boom: some are essential, such as resource-rich geology; others are catalysts, such as favorable land and mineral laws, which spurred the production of shale gas in the United States and Canada.[13]

Thanks to the recent glut in shale gas, the United States has started to export natural gas abroad. Now the export of technical and financial know-how is beginning to follow suit, though it will take other nations years to build up the infrastructure necessary for efficient shale oil and gas programs of their own.

Many countries are building pilot programs and experimenting with hydrofracking methods. Western Europe, the US EIA estimates, has some 639 trillion cubic feet of shale gas resources, which is more than four times the reserves of the Marcellus Shale.[14] Hydraulic fracturing was conducted in Germany, Holland, and the United Kingdom in the 1980s. In 2012, Cuadrilla Resources, a UK shale-gas explorer, calculated that its license area in Lancashire contains some 200 trillion cubic feet of natural gas, and estimates there is more shale gas in the UK than the entire reserves of Iraq.[15] (The UK uses roughly 3 trillion cubic feet of gas per year.) As Italy, Poland, and Lithuania build terminals to receive liquefied natural gas (LNG) tankers—long ships outfitted with distinctive spherical tanks—imports to the EU are estimated to grow by 74 percent by 2035.[16]

Eastern Europe's interest in hydrofracking has a political subtext. The first hydraulic proppant fracturing was carried out in 1952 in the Soviet Union, while Ukraine, Poland, and Mongolia are planning to tap shale basins, both for economic reasons and as a way to reduce their energy dependence on Russia, with whom they have had a fraught history.[17]

As for Russia itself, Vladimir Putin is eyeing America's shale boom with a mixture of deep suspicion and growing interest. Energy companies account for half the value of the Russian stock market, and their revenues help to prop up the government. Gazprom, the state-backed energy company with a monopoly on natural gas exports, has long used its dominance to get its way: Russia cut off natural gas to Ukraine twice, in 2006 and 2009, when contract negotiations grew heated. But as America uses its shale reserves to become the world's biggest gas producer and a likely gas exporter, the balance of power is shifting. America's gas glut is depressing prices on the world market, and sending unneeded gas from the Middle East to Europe. Bulgaria recently negotiated a 20 percent price cut in its new 10-year contract with Russia.[18]

Alexey Miller, the Gazprom boss, says, "We are skeptical about shale gas," and has derided the gas boom as a "myth" and "a bubble." But Gazprom is in trouble: its 2008 market capitalization of $367 billion had shrunk to a mere $78 billion five years later.[19] Putin—who once complained that shale gas costs too much and that fracking harms the environment—now says that there might be "a real shale revolution" after all, and that Russian energy firms must "rise to the challenge." Rosneft, the country's largest oil producer, is working to double its share of the Russian gas market by 2020, and might even try hydrofracking shale beds. As it turns out, geologists believe that Russia has enormous shale oil and gas reserves, which could one day supply both European and Asian markets.

In Asia, China is the world's biggest energy user and is estimated to have as much as 25 trillion cubic meters of shale gas reserves, 50 percent larger than US reserves.[20] China established a research center in 2010, drilled its first horizontal shale well in 2011, and has targeted the production of 30 billion cubic meters a year from shale. India is also thought to have significant shale gas reserves, far more in fact than the 38 trillion cubic feet that the EIA has officially estimated. In a 2013 speech at Pace University, Veerapa Moily, India's minister of petroleum and natural gas, described how his government is planning an aggressive push to extract shale gas from its large deposits.[21]

As in the United States, hydrofracking has not always been greeted with open arms elsewhere. France—an enthusiastic supporter of nuclear power, another controversial energy source—sits above the largest estimated reserve in Western Europe. But in 2011, after winemakers and environmentalists concerned about water and pollution pressured the government, France became the first nation to ban hydraulic fracturing outright.[22] In January 2012 Bulgaria joined the French, withdrawing a hydrofracking license granted to Chevron Corporation, after hundreds of protestors marched to the capital, Sofia, because of concerns about water and soil pollution in the nation's most fertile farming region, Dobrudja.[23] Other countries have placed a temporary moratorium on the practice, such as Czechoslovakia and Romania, pending legislation that eliminates legal risks. The United Kingdom also temporarily suspended hydrofracking, after two small earthquakes rocked Lancashire in northwestern England, but the ban was lifted in 2012.[24]

The lack of fracking in other countries is not simply due to public concerns. In Western Europe, for example, there is so little geological data about shale regions that in many cases it remains unclear how much gas is there and whether it can be extracted profitably. In 2012, ExxonMobile canceled plans to frack in Poland, after two test wells proved to be uneconomical.[25] Chevron also has a stake in Poland, but is proceeding "very cautiously."

In mid-2013 Germany was researching and debating the merits of hydrofracking. Peter Altmaier, Germany's environment minister, says the process should be banned near drinking water supplies. Yet Germany also has plans to phase out nuclear power generators completely, a decision that is sure to make natural gas "a bigger part of the energy mix," says Daniel Yergin, the noted American energy analyst.[26]

PART II

HYDROFRACKING

THE DEBATE

Like the public meeting I attended in New York City, the debate over hydrofracking usually devolves into acrimony, and it is difficult for ordinary citizens to know whether to support or oppose hydrofracking. As I explained in Part I, proponents claim that increased production of natural gas is leading to lower energy prices, more jobs, energy independence, and a reduction of greenhouse gases. Opponents say that the industry's rush to exploit shale reserves has led to water, air, and soil pollution and exposed communities to health problems, exploding wells, and earthquakes. The result is uncertainty and mistrust on both sides, and a certain amount of disinformation. In this part I take a close look at the pros and cons of hydrofracking, to help readers make informed decisions about an energy technology that is here to stay, for better or for worse.

5

THE CASE FOR HYDROFRACKING

Who Benefits?

Many have profited financially from hydrofracking, from George Mitchell, who made billions of dollars by inventing "slickwater" fracturing, to truck drivers in North Dakota, who saw their earnings jump to $2,000 a week hauling fracking fluid, to rural landowners in New York State, who leased their land for $5,000 to $6,000 an acre with 20 percent royalties, which could translate into hundreds of thousands of dollars per year.[1]

Or consider Terry Pegulia, who founded a company called East Resources with $7,500 borrowed from friends and family. Starting with one well in the Marcellus Shale region of western Pennsylvania, Pegulia eventually got the assistance of Ralph Eads III, a banker and a legendary salesman and a vice chairman at Jefferies & Company in Houston.[2] They parlayed that one well into 75 wells in a single year. In 2010, Pegulia sold East Resources to Royal Dutch Shell for $4.7 billion.

Floyd Wilson, a Texan who created Petrohawk Energy in 2003, drilled his first well in the Haynesville Shale in 2008, and sold the company to BHP Billiton, an Australian energy firm, in 2011 for $15 billion. Wilson and his executives earned $304 million.[3]

So for a lucky few, hydrofracking has been a highly lucrative gamble. Not everyone profits and the business remains volatile, yet the tales of quick riches are one of the things that keep others so feverishly interested.

What Is the Impact of Shale Oil and Gas on the US Economy?

In a word: revolutionary. As proponents point out, hydrofracking has helped to create new jobs, higher incomes, and tax windfalls for cash-strapped states.

For decades, the dialogue about oil and gas has been built on a paradigm of limited supply, declining production, and volatile pricing. These assumptions were based on the pioneering work of M. King Hubbert, a Shell geologist who in the 1950s forecast that American oil production would peak in 1971.[4] Shale gas reserves have turned those assumptions on their head.

The United States produced less oil in 2012 than in 1971—Hubbert was correct—and prices remained high. Yet America produced more oil in 2012 than in any year since 1994, and natural gas production is nearing record levels. In 2000, shale gas represented merely 2 percent of the nation's energy supply; by 2012, it was 37 percent.[5]

The abundance of gas has set off ripples throughout the US economy, with numerous additional impacts. As power companies substitute gas for coal in their generators, for instance, consumers benefit from lower prices. This switch has led to another important boon: a significant reduction in CO_2 emissions.

"One thing is clear," writes Steven Mufson, an energy and financial news reporter at the *Washington Post*. "Tumbling natural gas prices have changed every calculation and assumption about the energy business."

How Many Jobs Does Hydrofracking Create?

The development of shale resources supported 600,000 jobs in 2010, a number that the American Petroleum Institute (API) reports is "constantly increasing."[6] Many economists believe that the boom in hydrofracked fuels—the "shale gale"—will continue to rev the economy for years to come, and could give the United States a competitive advantage in the global marketplace.[7] Ed Morse, a Citigroup energy analyst, predicts that by 2020 the natural gas industry will have created some three million new American jobs. Hydrofracking, he says, will add up to 3 percent to America's gross domestic product and trillions of dollars of tax revenue.[8] In his 2012 State of the Union address, President Obama approvingly cited an estimate by the natural gas industry that hydrofracking will supply a century's worth of gas reserves and 600,000 new jobs by 2030.[9]

If the hydrofracking boom in the Marcellus Shale is any indication, he could be right. The Marcellus remains relatively untapped, but in 2009 hydrofracking there created over 44,000 new jobs in Pennsylvania, $389 million in state and local tax revenue, and nearly $4 billion in value added to the state's economy.[10] In the same year in West Virginia, which also sits atop the Marcellus, fracking added 13,000 new jobs and over $220 million in federal, state, and local tax revenue, and nearly $1 billion in value added to the state economy.[11]

How Has Supply Affected the Price of Natural Gas?

In 2010, proved reserves of natural gas and oil in the United States rose by the highest amounts ever recorded, according to the US Energy Information Administration: 2.9 billion barrels of crude oil (surpassing 2009's high by 63 percent), and 33.8 trillion cubic feet of natural gas (17 percent higher than 2009).[12] Largely thanks to hydrofracking, annual production has climbed to some 30 trillion cubic feet a year, and prices have dropped precipitously, from about $13 per million BTUs in 2006 to less than $4 per million BTUs in 2013.[13]

In short, abundant supplies have led to lower prices. America's Natural Gas Association (ANGA) estimates that between 2012 and 2015 low gas prices will add $926 in disposable income to each household, a number that could reach $2,000 by 2035.[14]

States like Pennsylvania, which was once a natural gas importer, are now producing so much shale gas that they are poised to export it.[15] Energy-intensive industries that have moved factories overseas in search of cheaper fuel are now able to stay in the States, or even "reshore" (bring back) facilities. Even in industries where electricity represents a minimal percentage of costs, the price of gas can have a profound impact.

The fossil fuel industry represents a modest slice of America's overall economic pie, but the relative drop in gas prices has been so dramatic that it could boost a manufacturing renaissance that would add 0.5 percent a year to GDP by 2017, according to the Swiss bank UBS.[16]

How Has Cheap Gas Impacted the Petrochemical Industry?

Industry uses around a third of US gas output, and one of the biggest consumers is petrochemical manufacturing.[17] Manufacturing products like methanol and ammonia, a key ingredient of fertilizer, require gas as feedstock (raw material for industrial processes). Switching feedstock from naphtha, derived from oil, to ethane, derived from gas, has kept petrochemicals affordable even as oil prices have spiked. These chemicals in turn provide cheaper raw materials for farmers, automakers, manufacturers of household products, and builders, or are exported at prices competitive with state-owned firms in the Middle East, the world's lowest-cost oil and gas producers.

In industries where petrochemicals are a large part of the cost base, companies could shorten the supply chain and return manufacturing work stateside. Should this come to pass, the accounting firm PricewaterhouseCoopers estimates, lower feedstock and energy costs could add one million new American factory jobs by 2025.[18]

The American Chemistry Council says that a 25 percent increase in the supply of ethane, a liquid derived from natural gas, could add over 400,000 jobs, provide over $4.4 billion in annual tax revenue, and add $16.2 billion in capital investment by the chemical industry alone.

How Have Chemical Companies Been Affected by Natural Gas Prices?

Louisiana provides a case study in how the fluctuating price of natural gas has shifted the fortunes of major corporations. Because the state is resource-rich, and natural gas once cost just $2 per thousand cubic feet, Louisiana enticed numerous chemical companies to build plants across the state. But in the fall of 2005 Hurricane Katrina devastated the Gulf Coast, and the price of natural gas jumped sevenfold, to $14 per thousand cubic feet. This sent a jolt through the industry. Yet even before the storm, Louisiana's gas supplies were dropping and prices were rising, which caused several big employers either to mothball plants or to move them.[19]

Consider the case of Dow Chemical. Gas represents half of its operating costs, and the skyrocketing prices after Katrina forced the temporary shuttering of one of its largest petrochemical plants. "The US is in a natural gas crisis," Dow CEO Andrew Liveris testified before the Senate at the time. "The hurricanes have dramatically underscored the problem, but they did not cause it....We simply cannot compete with the rest of the world at these prices." Switching its focus abroad, Dow invested in China and the Middle East—including a $20 billion joint venture to develop oil and gas facilities in Saudi Arabia—where energy prices were much lower. "Our industry will continue to grow," Liveris maintained. "It's simply a question of where we will grow."[20]

By the fall of 2012, however, natural gas prices had fallen dramatically, to about $4 per thousand cubic feet, and chemical industry employment figures nudged up for the first time in a decade, according to the Bureau of Labor Statistics. Dow Chemical restarted its plant in St. Charles, Louisiana, and planned to build a new complex in Freeport, Texas.

That same year, Methanex Corporation, the world's largest methanol producer—which had closed its last US chemical plant in 1999—paid over $500 million to disassemble a methanol plant in Chile and move it to Ascension Parish in Louisiana. "The proliferation of shale gas in North America…underpins the very attractive economics for this project," observed CEO Bruce Aitken. If gas prices remain steady, he added, the Chilean plant will be paid off by 2016, and the company might move a second plant from Chile to Louisiana.[21]

Similarly, the Williams petrochemical company was spurred by low gas prices to invest $400 million in expanding an ethylene plant. And CF Industries announced a $2.1 billion expansion of a nitrogen fertilizer plant. Gas prices account for 70 percent of its manufacturing costs at its ammonia and urea units (urea is used in fertilizers and to synthesize resins and plastics).[22] With cheap natural gas, the company is betting it can successfully compete against imported nitrogen fertilizers, which represent over half of sales in the United States.

What Is the "Halo Effect" of Gas Prices on Other Industries?

This kind of growth is already having a beneficial impact on numerous industries associated with energy, from trucking to high tech, glass, sand, steel, and even plastic toys.

U.S. Steel, for instance, has struggled in recent decades, undercut by Asian producers and unable to respond creatively, and saw its stock price drop nearly 90 percent. But today it is in the midst of a Phoenix-like resurrection. Hydrofracking requires lots of steel products, from pipelines to drilling rigs and well casings, and U.S. Steel recently invested $100 million on a facility to produce "tubular product" specifically for the hydrofracking industry. The company may see even bigger gains in coming years, as the shift from coal to cheaper natural gas reduces its energy costs. In 2011, U.S. Steel used 100 billion cubic feet of natural gas: every time the price dropped a dollar, the company saved $100 million. In a further cost-saving measure, the company is retooling its blast furnaces to reduce the amount of coke (a fuel derived from coal) used, and increase the amount of natural gas it can inject. The result is that blast-furnace costs could be cut by as much as $15 per ton of hot metal produced. When demand is strong, U.S. Steel produces some 20 million tons of steel annually. The company's advantage grows as competitors in Asia and Europe are forced to pay more for their energy. "It has become clear to me that the responsible development of our nation's extensive recoverable oil and natural gas resources has the potential to be the once-in-a-lifetime economic engine that coal was nearly 200 years ago," announced U.S. Steel chairman John Surma in a 2012 speech.[23]

As the tide of natural gas rises, other industries are seeing their boats rise with it. One might not associate high-tech companies with natural gas; yet a few of them have leaped on the shale bandwagon. In 2012, Honeywell—known for making thermostats, electric motors, and components for nuclear weapons—paid $525 million for a majority stake in Thomas Russell, a provider of equipment for natural gas processing and treatment. In coming years, Honeywell will offer products that allow energy companies to remove contaminants from hydrofracked natural gas, and to recover natural gas liquids used for fuel and in petrochemicals.[24]

As energy costs dip, companies along the value chain (a series of companies that work together to deliver products or services to the market) are retrenching; some that fled the United States for foreign shores are contemplating a return home. The manufacture of plastic toys by injection molding, for instance, requires lots of power but minimal labor costs.

For now, Chinese factories dominate the market. However, given the costs of transportation, lengthy supply chains, and other factors—such as fear of intellectual property theft—the calculation about where manufacturing takes place is shifting.

What Are the Nonindustrial Benefits of Hydrofracking?

According to a study done by scientists at MIT, residential and commercial buildings account for 40 percent of America's total energy consumption, in the form of electricity or gas, making up over half the country's demand for gas.[25] Low gas prices have meant that the cost of heating schools and other government buildings, often itemized on local tax bills, is falling.

How Has the Natural Gas Bonanza Affected Foreign Investment in the United States?

After years of losing manufacturing jobs, communities near productive shale plays are using incentives—and their proximity to natural gas supplies—to lure foreign investments.

In 2012, for example, Orascom Construction Industries, based in Cairo, Egypt, and one of the world's biggest fertilizer manufacturers, announced it would build a $1.4 billion plant in Wever, Iowa. The company chose this site over one in Illinois because part of its investment will be funded by a tax-exempt bond that provides $100 million in tax relief. The plant, Orascom announced, will be "the first world-scale, natural gas-based fertilizer plant built in the United States in nearly 25 years."[26]

A number of states that sit atop the Marcellus Shale—Ohio, West Virginia, and Pennsylvania—were recently engaged in a head-to-head competition to woo Royal Dutch Shell, the energy firm based in London and The Hague. The company ultimately decided to build a $2 billion petrochemical plant northwest of Pittsburgh.[27]

How Do New Shale-Gas Supplies Affect the Global Energy Market?

Currently the 12 member nations of OPEC produce over 40 percent of the world's oil, which gives the group tremendous control over the price of crude oil, the biggest factor in gasoline costs.[28] As the United States adds to its energy reserves through hydrofracking shale, OPEC's influence on prices at the pump will weaken.

Indeed, a 2012 report by the National Intelligence Council (NIC), an adviser to the CIA, found that the success of American shale oil and gas exploration could soon cause a fundamental shift in the global energy market. The NIC estimates that US oil production could expand to 15 million barrels a day, more than double the current rate. This would reduce domestic oil prices, increase US economic activity by 2 percent, add 3 million jobs, and could turn the United States into a major oil exporter by 2020.[29] A separate Energy Department analysis found that the United States now has sufficient natural gas supplies to make it a major exporter. "In a tectonic shift, energy independence is not unrealistic for the US in as short a period as 10–20 years," the NIC found.[30]

Another report, "Oil: The Next Revolution," compiled by the Belfer Center at Harvard, confirms this prediction: shale oil could provide as much as 6 million barrels a day by 2020, which would bring the United States close to energy independence in oil by then.[31]

Alan Riley, professor of energy law at the City Law School at City University London, notes: "The major geopolitical impact of shale extraction technology lies less in the fact that America will be more energy self-sufficient than in the consequent displacement of world oil markets by a sharp reduction in US imports."[32]

As shale oil supplies are tapped in China, India, England, Australia, Argentina, and elsewhere, global oil prices will sink further—a possibility that was unthinkable just a few years ago.

How Will This Affect Transportation?

The place where natural gas increases could ultimately have the biggest impact is by replacing gasoline in the world's cars, trucks, and buses. At the moment, transportation accounts for 70 percent of America's petroleum use, and 30 percent of US carbon emissions.[33] In 2011, President Obama vowed to cut oil consumption by a third in the next decade, and two years later he said that emissions would be cut by 17 percent by 2020.[34] Natural gas is an important part of his plan to reduce oil imports and greenhouse gases.

The combustion of natural gas produces 30 percent less carbon dioxide than oil, and could be used instead—either directly, as compressed natural gas (CNG) or liquid natural gas (LNG), or indirectly, by converting natural gas into liquid fuel or power for electric vehicles.[35] At the moment, just 1.5 percent of the world's cars and trucks—about 16 million vehicles—are powered by natural gas. Morgan Stanley estimates that the gas used by these automobiles replaces about 1.2 million barrels per day of oil products.[36] In a best-case scenario, that use could rise to 5.6 million barrels per day, or 7 percent of oil supplies, in a decade.

The United States and Western Europe already have a gas infrastructure that pipes gas into buildings that is well developed; the so-called "last mile problem" (a phrase borrowed from the telecom industry, which refers to the final leg of a delivery system that connects suppliers with customers) can be solved with devices that allow people to refuel their natural gas-powered vehicles (NGVs) at home.[37] This is not as far-fetched as it might sound. The number of NGVs in the United States doubled between 2003 and 2009 to 110,000. That is a miniscule number, representing only 0.1 percent of all vehicles on the road, but it is growing.

The Dallas–Fort Worth Airport runs 500 maintenance vehicles on gas (and allows hydrofracking beneath its runways). AT&T is buying 8,000 CNG-powered vehicles, giving it the largest commercial NGV fleet in the country. School buses, garbage trucks, and other municipal vehicles are switching.[38]

CNG is not ideal. It has to be stored at high pressure in bulky tanks. An average-sized CNG tank provides only a quarter of the traveling distance of gasoline tanks. Retrofitting vehicles with CNG equipment is costly, and refueling infrastructure is not widely available. There are only 1,500 public CNG stations in the United States, compared with 115,000 regular filling stations. Still, CNG is catching on among fleet-delivery vehicles (FedEx and UPS trucks, for instance) and public buses. Some 20 percent of local buses already run on CNG or LNG.[39]

Natural gas could also power ships. In an effort to save fuel and reduce emissions, ferries in Argentina, Uruguay, Finland, and Sweden are already powered by LNG. In 2014, a new LNG ferry will start plying the St. Lawrence River in Quebec. And in a New York City pilot project, one of the Staten Island ferries is being retrofitted from a low-sulfur diesel to an LNG power plant; the switch will cut fuel consumption in half and reduce greenhouse gas emissions by 25 percent. Such pilot programs could lead to similar conversions on much bigger, oceangoing container ships. The research firm IHS CERA projects that by 2030 a third of all cargo carriers will be fueled by LNG.[40]

In Europe, regulators are pressuring shippers to reduce emissions, especially on inland waterways. Royal Dutch Shell has chartered two LNG-powered tanker barges to work on the Rhine River. The barges emit 25 percent less carbon dioxide than their oil-burning competitors, and are quieter.[41]

Another way to fuel transport is through gas-to-liquids (GTL) technology, which uses heat and chemistry to convert natural gas into liquid fuel (similar to the way crude oil is converted into gasoline). The technology uses catalysts to turn gas into longer-chained hydrocarbons, like diesel and kerosene, or various petrochemicals. There are now several GTL plants operating around the world. Shell's $19 billion Pearl facility in Qatar (jointly owned by the Qataris) is by far the largest, and Shell may build a similar facility on the Gulf of Mexico.[42]

What Impact Has Hydrofracking Had on Water Supplies?

A 2011 MIT report coauthored by Ernest Moniz, President Obama's new secretary of energy, found that natural gas exploration has, overall, had a good environmental record.[43] This gladdened the industry, of course, which maintains that hydrofracking takes place thousands of feet below the water table, and the chemicals it uses are typically separated from groundwater by impermeable rock. Supporters note that over 20,000 wells have been drilled in the past decade, and that only a few instances of groundwater contamination have been reported, all of them due to breaches of existing regulations. "The data show the vast majority of natural gas development projects are safe, and the existing environmental concerns are largely preventable," Scott McNally, an environmental engineer who has worked for Shell, blogged for *Scientific American*.[44]

In 2011, then-EPA administrator Lisa Jackson testified to the Senate that she was unaware of "any proven case where the fracturing process itself affected water."[45] In 2012, officials from the Department of the Interior told Congress that "we have not seen any impacts to groundwater as a result of hydraulic fracturing."[46]

An average shale well uses a lot of water—some 1.2 to 5 million gallons over its lifetime[47]—but, the industry says, that number is not as alarming as it might sound. To put it in perspective, all of the shale wells drilled and completed in 2011 used 135 billion gallons of water, which was equivalent to 0.3 percent of total US freshwater consumption, according to an analysis by TheEnergyCollective.com (an independent forum supported by Siemens and Royal Dutch Shell).[48] Yet agriculture used 32,840 billion gallons of water annually (243 times more water than hydrofracking for natural gas), and golf courses used about 0.5 percent of US fresh water.

Furthermore, the volume of water used by hydrofracking compares favorably to that used to produce other forms of energy. According to ConocoPhilips, natural gas uses about

60 percent less water than coal and 75 percent less water than nuclear power generation.[49]

What Is the Halliburton Loophole, and How Do Drillers Respond to the Charge That It Conceals the Chemicals Used in Hydrofracking?

When hydrofracking opponents complain that public health is put at risk by the 2005 ruling under which companies are not required to disclose some of the chemicals they use because they are trade secrets, energy groups vehemently deny they are hiding anything nefarious. This special exemption for hydrofracking, reportedly inserted by Vice President Dick Cheney, a former CEO of Halliburton, was dubbed "the Halliburton Loophole" by opponents.[50]

One industry group, Energy in Depth (EID), maintains that state regulations governing hydrofracking fluids are robust, and that there is no need for federal oversight. "Getting the public to believe that hydraulic fracturing is essentially unregulated is critical to some folks' strategy of shutting it down," EID writes. "But here's the truth: States have regulated the fracturing process for more than six decades now, and by any legitimate measure have complied an impressive record of enforcement in that time…(hydrofracking) has never in its nearly 65-year history been regulated under the Safe Drinking Water Act.…If a bill never covered you in the first place, how can you be considered 'exempt'? Does that mean [hydrofracking companies] are exempt from Medicare Part D too?"[51]

EID notes that "there isn't a single 'hazardous' additive used in the fracturing process that's hidden from public view," adding, "So what's with all the controversy over 'trade secrets'? In rare cases, a company may ask that a certain 'constituent' contained within a larger 'additive' set be protected, though…under law that information must be released to response and medical personnel in case of emergency.…Indeed, the vast majority of these (chemical constituents) are considered 'non-hazardous' by EPA—quite the contrast from what you've read in the papers."

As we will see in the next chapter, there are those who strongly disagree with this assertion.

How Has Hydrofracking Affected Global Warming?

This question, perhaps more than any other, epitomizes the public confusion and sharp disagreement that splits the two sides of the debate over hydraulic fracturing. Adding to the misunderstanding, regulators at the federal EPA have shifted their position on this question.

One of the main justifications for promoting natural gas is that power plants fueled by gas emit about half the climate-changing gases that coal-fired plants do.[52] Advocates say the gas boom has been a key reason the United States is the only major nation to see significant reductions in climate-warming emissions. But some argue that methane leaks in gas operations nullify those benefits, because methane (the primary component of natural gas) is a formidable greenhouse gas.

First the good news. Between 2007 and 2012 the United States reduced its greenhouse gas emissions by 450 million tons, the biggest drop of any nation in the world.[53] Hydrofracking advocates are quick to claim that this is a result of the nation's fundamental shift from coal to natural gas and renewable energy.

In 2000, coal powered 52 percent of US electric generation, natural gas provided 16 percent, and renewables were only 9 percent. By 2012, according to the Energy Information Administration, coal had plummeted to 38 percent, gas had jumped to 30 percent, and renewables had risen to 12 percent.[54] "Pragmatic leaders understand...that with natural gas we don't have to choose between our economic and environmental priorities. Instead, responsible natural gas development is having very real and positive impacts in the 32 states that are home to this abundant domestic energy source," the American Natural Gas Alliance writes in a recent report.[55]

A study by the environmental investor group CERES found a nearly 70 percent reduction in sulfur dioxide and smog-forming nitrogen oxide over the past 20 years, thanks to growing use of natural gas among the nation's top 100 utilities.[56] Nearly half the reduction came in just a two-year period from 2008 to 2010. The EIA has similarly noted that US carbon emissions from the power sector are at 20-year lows, largely due to the increasingly prominent role of natural gas in the nation's energy portfolio.[57] (The European Union, by contrast, has seen its greenhouse gas emissions rise over the same time-frame, despite a more concerted effort to tackle global warming than the United States has made.[58] This is driven by the EU's increased reliance on coal for generating power.)

However, methane is the main component of natural gas, and it has a notoriously potent greenhouse effect, with a global warming potential 72 times higher than carbon dioxide—the leading greenhouse gas—over a 20-year period, according to the Intergovernmental Panel on Climate Change, and 20 times greater over a 100-year period, according to the EPA.[59] The EPA has long held that natural gas operations represent the leading source of methane leaks in the United States, accounting for 145 metric tons in 2011. (The second-largest source of methane was "enteric fermentation," aka gas emitted by cows and other animals, at 137 metric tons. The third-largest source were landfills, which emitted 103 metric tons.)[60] Yet the EPA estimates that all sources of methane combined represent just 9 percent of greenhouse gases.

The industry has been working hard to reduce emissions, in part to build public acceptance of hydrofracking but also because every methane leak represent a loss of lucrative product.

In April 2013, the EPA dropped a bombshell of sorts, announcing that it had dramatically reduced its estimate of the amount of methane the gas industry leaks. Although gas production has grown 40 percent since 1990, thanks to hydro-fracking, the EPA found that tighter pollution controls—from better equipment, maintenance, and monitoring—resulted in an average decrease of 41.6 million metric tons of methane annually between 1990 and 2010, or over 850 million metric tons in total.[61] The EPA's new numbers represent a 20 percent reduction from the agency's previous estimates.

Both advocates and opponents erupted at the EPA announcement. "The methane 'leak' claim just got a lot more difficult for opponents," blogged Steve Everly, of industry group Energy in Depth.[62] But Robert Howarth, a professor of ecology at Cornell University who authored a celebrated methane study critical of hydrofracking, flatly stated, "The EPA is wrong." Howarth and researchers at the National Oceanic and Atmospheric Administration (NOAA) had recently published a new study detailing huge methane leaks from gas-drilling sites in Colorado and other western states. He said the EPA was "ignoring the published NOAA data," and called for "an independent review of (EPA's) process."[63]

A leading environmentalist says the latest fracas misses the larger point. "We need a dramatic shift off carbon-based fuel: coal, oil and also gas," Bill McKibben, founder of the climate group 350.org, told the Associated Press. "Natural gas provides at best a kind of fad diet, where a danger-ously overweight patient loses a few pounds and then their weight stabilizes; instead, we need at this point a crash diet, difficult to do" but necessary to limit the impact of global warming.[64]

The EPA says that despite its revised estimate, natural gas operations remain the country's leading cause of meth-ane emissions. The agency has vowed to continue collect-ing data and researching the subject, and may change its conclusions again.

The acquisition of shale gas is depicted by the energy indus-try as an unquestionably good idea in a time of economic uncertainty: a vast, clean, 100-year supply of energy. In coming years, hundreds of thousands of new wells across the United States and, perhaps, around the world will be hydrofractured. But the true environmental and health costs of this method of extracting natural gas and oil are not well understood, and have many—even some who support hydrofracking—concerned.

6

THE CASE AGAINST HYDROFRACKING

Hydrofracturing is not a gentle process. Sucking oil and gas from dense shale formations involves drilling, explosions, toxic chemicals, and millions of gallons of water pumped at crushing pressures. Drillers maintain that these processes are well understood and tightly controlled and take place far below groundwater supplies. But ultimately the safety and quality of a well is dependent on the operator, the particularities of each site, local regulations and politics, and many other details that can get lost amid the chaos of a drill pad. As the shale revolution has gained momentum, it has provoked an increasingly vocal backlash, with protestors from Grand Rapids to Paris calling for a "global frackdown."[1] In the United States, people worry that in the rush to embrace shale energy Congress granted hydrofrackers special exemptions from federal regulations—the Clean Air Act, the Clean Water Act, and the Safe Drinking Water Act—without thinking through the potential health and environmental consequences.

While most hydrofracturing has been conducted responsibly, the industry does not have a perfect track record: from time to time, gas wells blow out, water supplies are poisoned, soil and air are polluted, and the health of people and animals is compromised. Protests over these incidents have inflamed "fracktivists."

The debate over the Keystone XL pipeline—which could bring tar sands oil from Alberta Canada to the Gulf Coast—has raised the specter of groundwater pollution by fracking. And a spate of accidents by large resource companies—the explosion of BP's *Deepwater Horizon*, which caused the worst offshore oil spill in US history; an oil leak in Brazil and a refinery fire in California by Chevron; the fumbled attempt to drill in the Arctic by Royal Dutch Shell; and a rupture in ExxonMobil's Pegasus pipeline that spilled crude oil in an Arkansas housing development—have stiffened opposition to shale exploration. Each of those companies is also engaged in hydrofracking.

Once a shale formation has been fracked, it cannot be unfracked and pieced back together again, opponents say. So it is prudent to ask tough questions and push for complete answers before greenlighting widespread energy exploration, especially in populated areas (such as the New York City watershed, above the Marcellus Shale). This line of thinking has created unlikely alliances between ranchers, industry, and environmentalists in places like Texas and Colorado. Forty years after the founding of the Environmental Protection Agency and signing of the Clean Water Act, the debate has reenergized the slumbering environmental movement, and attracted celebrity movie stars and musicians to the cause. In their 2012 single "Doom and Gloom" the Rolling Stones lament: "Fracking deep for oil, but there's nothing in the sump....I'm running out of water, so I'd better prime the pump!"

So deep is the divide between advocates and opponents that a straightforward conversation about hydrofracking is nearly impossible in certain communities. The stalemate has some advocates worried that resistance could hobble the shale revolution. The IEA notes that "concerns remain that production might involve unacceptable social and environmental damage," and recommends that in order to preserve the "gas revolution" drillers engage with their opponents, be transparent about the chemicals and processes used, and submit to tighter regulations for the greater good (more on this in chapter 7).[2]

Yet ideally the questions skeptics ask should help improve communication and bridge the divide between the two camps, demystify hydrofracking, protect health and investments, and forestall the kind of environmental debacle that could set the entire industry back.

What Questions about Hydrofracking Need to Be Asked and Answered, According to Opponents?

These questions begin with the hydrofracturing process itself, and the steps that are involved, as outlined earlier: the construction and operation of the drill pad; the drilling, integrity, and performance of the borehole; the injection of fluids underground; the flowback of these liquids to the surface; the capture, processing, and transportation of oil or gas; the disposal of wastewater; and the eventual closing down of the well.

Each of these steps raises concerns about various issues, including the depletion or pollution of water supplies; the mishandling of chemicals and waste at the surface; well blowouts; exposure to naturally-occurring radioactive nuclides; the migration of gas or other fumes into the air; contamination of food supplies; adverse health effects in man and animals; and earthquakes caused by the injection of wastewater into fault zones. Less quantifiable but still significant is the social impact of fracking—that is, what happens when rural landowners become rich, or don't, by leasing their property to frackers; the impact on small communities when thousands of roughnecks suddenly appear to frack wells, then just as suddenly leave when the job is done; or the consequences of having hundreds of big trucks rumbling on country roads and dozens of noisy, brightly lit drill rigs operating 24/7.

What Are the Biggest Concerns in Terms of Water Supplies?

As we've seen, water constitutes the largest component of fracking fluid by far, so it is not surprising that questions about the quantity and quality of water used by drillers have been contentious.

Opponents have expressed three main concerns about water. First, they worry that hydraulic fracturing uses so much H_2O—about 5 million gallons per well, on average—that it can deplete groundwater supplies faster than nature can recharge them, especially in dry regions like Texas or California.[3] ("Recharge" signifies the amount of water an aquifer—an underground water supply—regains each year from precipitation and runoff.) Second, the injection of chemicals—some of them toxic—underground at extreme pressures raises fears of chemical spills on the surface and consequent contamination of water supplies below ground as those chemicals seep into the fractured rock. Third, the safe disposal of fluid and "produced water" (groundwater that is brought to the surface in the course of drilling) remains a challenge, and has occasionally caused minor earthquakes when injected into geologic fault zones.

Because hydrofracking is new to many regions, legislators and regulators are scrambling to catch up with these issues—with, as we shall see, varying degrees of success.

Does Fracking Deplete Aquifers?

While the amount of water used in hydraulic fracturing depends on the type, depth, location, and characteristics of each shale formation, a typical drilling operation will use 6,000 to 600,000 gallons of fluids just in the initial stages of the process, according to Chesapeake Energy.[4] Over the course of their lifetime, some wells will use 2 to 4 million gallons, though others—such as those in Texas' Eagle Ford shale—can use as much as 13 million gallons.[5] (It should be noted that there are many variations among shale formations, that estimating water use is complex, and because no fractured well has yet experienced an entire life cycle these numbers are well-educated guesses.)

Water underlies most other resources, and as hydrofracking spreads it has set off "resource wars"—a competition for limited water supplies—that, in a state like Colorado, pit energy companies against traditional users such as farmers, ranchers, builders, industry, ski areas, and homeowners.[6]

In Texas, where the population is growing and a brutal drought has lingered since 2010, a study by the University of Texas found that the amount of water used in hydrofracturing more than doubled between 2008 and 2011.[7] This amount will likely increase before leveling off at about 125,000 acre-feet in the 2020s. In 2011, 632 million barrels of water were used to produce 441 million barrels of oil. Some studies show that hydrofracking consumes less than 1 percent of the total water used statewide, which is much less than agriculture or even lawn watering in Texas. But water tends to be a local issue, and in drilling hotspots like Dimmit County, water use has grown by double digits to keep pace with the shale oil boom.

Luke Metzger, director of Environment Texas, charges that the industry is "absolutely not doing enough" to conserve water.[8] Legislators have convened hearings and undertaken studies, and have pushed the industry to conserve. But in 2011, only one-fifth of the water used for hydrofracking was brackish or recycled water; the rest was clean water. Furthermore, hauling water to well pads and taking wastewater away requires hundreds of trips by heavy trucks, which adds traffic, wears out roadways, and antagonizes the public. Oil and gas drillers are "in the spotlight right now," said state representative James Keffer, the Republican chairman of the Texas House Energy Resources Committee. "They have to prove themselves."[9]

As hydrofracking technology spreads around the world, concerns like these will follow it. According to the Oxford Institute for Energy Studies, European shale formations lie 1.5 times deeper than those in the United States, and require more fluid to hydrofrack.[10] And in hot, dry nations with fast-growing populations—such as India, Australia, and South

Africa—water is already a grave concern; hydrofracking will therefore add further competition for limited supplies.

Does Hydrofracking Contaminate Groundwater?

Oil- and gas-bearing shale formations deep underground are often connected by cracks, fissures, and channels to water-bearing formations. The latter hold groundwater, and many worry that chemicals, seeping methane, and other pollutants will contaminate people's drinking supplies.

Even in the best-run hydrofracturing operations, there are many opportunities for water pollution, a risk that increases significantly with wildcat operators who sometimes manage their drill sites less carefully than well-established firms.

Engineers who investigate industrial accidents note that as equipment and industrial processes grow increasingly sophisticated, and reach deeper and deeper into the earth, the "human factor" often leads to costly mistakes.[11] Bad judgment can lead to well blowouts. Poorly built or damaged, boreholes and pipelines—such as those used to transport wastewater to treatment plants—can allow pollutants to flow into groundwater. Accidents involving trucks or the storage of fracking chemicals can lead to chemical spills, and the runoff will eventually be flushed into rivers, streams, or aquifers. Wastewater storage ponds allow volatile compounds—such as benzene, xylene, and naphthalene—to evaporate into the atmosphere, and can overflow when it rains.[12]

Hydrofrackers say they use an array of sophisticated engineering techniques—such as magnetic resonance imaging and sonar—to study their underground explosions and carefully control the extent of the fractures in shale formations, and thus the spread of fluids.[13] They say that the actual fracturing happens thousands of feet from water supplies and below layers of impenetrable rock that seals the world above from what happens down below. Yet this is not always the case. Even if freshwater supplies are sealed off from the region where fluid is injected, the gas well itself can create openings in rock: a borehole is surrounded by cement, but often there are large empty pockets, which can cause buckling, or the cement itself can crack under pressure. The powerful pumps can cause gas and fluids to leak into surrounding water supplies.[14]

In the United States, hydrofracking is suspected in at least 36 cases of groundwater contamination, and in several cases EPA has determined that it was the likely source of pollution.[15]

A report by the Ground Water Protection Council found that only 4 of the 31 drilling states it surveyed have regulations that address fracturing, and that no state requires companies to track the volume of chemicals left underground.[16] One in five states doesn't require that the concrete casing used in wells be tested before hydrofracking begins. And more than half the states allow waste pits filled with fluids to intersect with the water table, even though the pits have allegedly caused water contamination.

Drillers have developed methods to reduce spills and seepage of chemicals, but are usually left to implement them on their own. The result is that protections at drilling sites just a few miles apart can be completely different.

In a 2011 conference call with reporters, Richard Ranger, an American Petroleum Institute senior policy adviser and frequent commentator on hydrofracking, said: "The issue of where do these fracking fluids go, the answer is based on the geology being drilled....You've got them trapped somewhere thousands of feet below with the only pathway out being the wellbore. I'm just not sure that that study is out there."[17] Aside from the startling admission that the industry doesn't know where fracking fluids end up, Ranger added that there is no way to conclusively determine whether hydrofracking is safe or unsafe.

Following up on this question, the independent online newsroom ProPublica queried over 40 academic experts, scientists, industry officials, and federal and state regulators. None of them could provide a definitive answer.[18]

How Well Regulated Is Groundwater?

Despite assurances from those in favor of fracking that there are no proven cases of affected water, numerous cases of suspected groundwater contamination have been documented, and science writer Valerie Brown, for one, predicts that "public exposure to the many chemicals involved in energy development is expected to increase over the next few years, with uncertain consequences."[19]

In December 2011 the Environmental Protection Agency released its first thorough study of groundwater pollution with a draft report on the drinking water in Pavillion, Wyoming, which contained "compounds likely associated with...hydraulic fracturing."[20] The multiyear study was peer reviewed by scientists, and was among the first by the government to directly link fracking with groundwater pollution. It was considered a "blockbuster" by fracking opponents.[21] But then the agency seemed to back off from its conclusions.

The EPA's investigation began in 2008, when Pavillion residents complained that their water had turned brown and undrinkable. The EPA drilled its own wells near hydrofracking operations, and in sampling the groundwater detected methane and "high concentrations of benzenes, xylenes, gasoline range organics, diesel range organics and...hydrocarbons in ground water samples...[and] water near-saturated in methane." Benzene was found in one well at concentrations of 246 micrograms per liter, far beyond the legal standard of 5 micrograms per liter.[22]

EPA scientists tried to find other potential sources for the pollution, but concluded that the organic compounds must have been "the result of direct mixing of hydraulic fracking fluids with ground water," and advised locals to stop drinking from their wells.[23]

EnCana Corporation, the Canadian company that drilled the wells, and North America's second-largest producer of natural gas (after ExxonMobil), denied its wells had polluted

Pavillion's water. A few energy-policy analysts agreed, saying the EPA's evidence was "incomplete." Nonetheless EnCana's shares dropped over 6 percent on the New York Stock Exchange, and the incident hit other companies, such as Chesapeake Energy Corporation, which fell 5.1 percent.[24]

Later, after complaints by industry and Republican legislators, the EPA softened its position on the Pavillion case and said it would await the results of a peer review of its science. In June 2013, the agency abruptly announced that it would discontinue the peer review and turn the study over to the state of Wyoming.[25] In what would appear to be a conflict of interest, the state's research will be funded by EnCana, the company at the center of the dispute. Industry boosters said the EPA's decision was simply a long-awaited recognition that the agency had overreached in Pavillion. Opponents, however, were aghast, and charged that the EPA was ducking its responsibilities. "The EPA just put a 'kick me' sign on it," blogged John Hanger, a Democratic gubernatorial candidate in Pennsylvania, and the former secretary of the state's Department of Environmental Protection. "Its critics from all quarters will now oblige."[26]

Indeed, fracking opponents detected an unsettling trend, according to ProPublica. In 2012 the EPA's budget was cut 17 percent, to below 1998 levels, while sequestration cuts starved research funds for cases like the one in Pavillion.[27] And the EPA seemed to be in retreat on numerous fronts: a probe into groundwater pollution in Dimock, Pennsylvania, was canceled; a claim that methane released by a driller in Parker County, Texas, was contaminating residents' tap water was dropped; and a 2010 estimate that showed gas leaks from wells and pipelines was contributing to climate change was sharply reduced.

"We are seeing a pattern that is of great concern," said Amy Mall, a senior policy analyst for the Natural Resources Defense Council. The EPA needs to "ensure that the public is getting a full scientific explanation."[28]

The agency has said that the series of decisions were unrelated, and that the Pavillion case could be handled more quickly by Wyoming officials. Yet, in private, EPA officials have acknowledged that brutal political and financial pressures are tying their hands when it comes to enforcing environmental protections.[29]

The EPA under President Obama has paid close attention to hydrofracking operations, but this may have collided with the president's plan—outlined in a major policy speech in 2013—to reduce greenhouse gas emissions by relying heavily on natural gas.[30] The Obama EPA, critics say, has not always rendered clear and consistent decisions.[31]

When a family in the Fort Worth suburb of Weatherford found their water bubbling "like champagne," suspicions fell on nearby hydrofracking by Range Resources, a leading independent natural gas driller. In late 2010, the EPA issued a rare emergency order declaring that at least two homes in Weatherford were in immediate danger from a well saturated with methane and benzene. EPA required Range to clean up the well and provide the homeowners with safe water. The state backed up Range, who denied the contamination had been caused by their drilling, but an independent investigator found that chemicals in the well were nearly identical to the gas that Range was producing. Then the dispute shifted into federal court just as the EPA was asking Range and other energy companies to participate in a national study of hydrofracking. Range declined to participate so long as the agency pursued its action in Texas. In March 2012, EPA retracted its emergency order in Weatherford, ended the court battle with Range, and refused to comment on the case other than to say it was moving on to focus on "a joint effort on the science and technology of energy extraction." The homeowner who brought the case was outraged, and critics charge that the EPA had "dropped the ball."[32]

As a consequence of the EPA's shifting stance, environmentalists and some energy analysts have not been impressed with the regulatory oversight of hydrofracking. But the agency insists it takes such contamination issues seriously. As of this writing, the EPA is in the midst of a major national study on the environmental impact of the drilling technique and will publish a draft report in 2014—though it has warned that its final results will not be made public until 2016, President Obama's last full year in office.[33]

What Is Methane Migration?

As we've seen, methane is the main component of natural gas, and it has a climate-changing potential 20 times greater than carbon dioxide when measured over a 100-year period. While methane is not toxic, if allowed to concentrate in an enclosed space it carries a high risk of combustion.

Hydrofracking opponents have been galvanized by suspicions that methane can migrate from wells into underground aquifers and water wells, and then into the atmosphere. They suspect that cement well casings are not always sound, and worry that groundwater quality is diminished by the gas. There have been a number of documented cases of methane migration. As early as 1987, the EPA reported that fluid from a gas well hydrofracked in Jackson County, West Virginia, contaminated a private well in 1984.[34] But the most famous case centers on the Appalachian town of Dimock, Pennsylvania (pop. 1,500). In 2006, a cadre of "landmen" appeared in Dimock and quickly convinced residents to sell their mineral rights for $25 an acre. (Similar deals in neighboring towns would later cost $4,000–$5,000 an acre.)[35] Two years later, the landscape was dotted by drilling towers and hydrofracking equipment, and soon Dimock was home to some of the most productive gas wells in the state. But residents' drinking supplies turned brown or orange, the water smelled sulfurous, and in at least one case methane built up in a private water well and exploded. When tested, local water showed dangerous levels of methane, iron, and aluminum.[36] Pets and farm animals shed hair. Sores appeared on the legs of children, and adults suffered from ringing headaches.[37]

The press took note, and Dimock became known as "ground zero" in the dispute over the safety of hydraulic fracturing.[38] The most notorious scene in the documentary film *GasLand* shows a Colorado man lighting his tap water on fire. It is a startling sight, and activists insist that it demonstrates how methane can migrate from hydrofracked wells into peoples' drinking supplies in places like Dimock. The industry disputes this, however, and says that naturally occurring (or "biogenic") methane had infiltrated Dimock's water long before the frackers arrived. The "flaming water" scene in *GasLand*, they say, is merely a parlor trick used to scare the public.[39] Some water wells do indeed descend through many layers of shale and coal, which can naturally seep methane into groundwater—by so-called "methane migration." But reports of methane migrating naturally into wells date back to the 1800s, and while such incursions of gas may or may not be related to drilling, there is no conclusive evidence that they are the result of fracking.[40] Even if investigators use isotope identification to nail down a particular well's unique gas "fingerprint," it is impossible to prove the methane migrated into a water supply unless the water had been tested before drilling began. But in an open letter to audience, press, and peers, *GasLand* director Josh Fox insists that he has his facts right.[41]

Regardless, high levels of chemicals associated with hydrofracking—arsenic, barium, DEHP, glycol compounds, manganese, phenol, and sodium—were found in the drinking water of Dimock. In 2009, 15 local families sued Cabot Oil and Gas, the Houston-based energy firm for allegedly tainting their wells. (Cabot had 130 drilling violations in Dimock.) Pennsylvania—a pro-hydrofracking state—fined Cabot $120,000, banned it from drilling further wells in Dimock (though existing ones were allowed to continue operations), and demanded the company provide clean drinking water to 10 households. A consent decree was signed, and in 2010

Cabot was reportedly ordered to pay a $4 million settlement.[42] Some families accepted methane treatment systems from the company, and, perhaps worried about their plummeting real-estate values, now insist the water is fine. But others, convinced that methane was only one of several chemicals tainting their water, sued Cabot.

In 2011, EPA informed Dimock residents that their well water was not an immediate health threat; but in January 2012, the agency reversed itself and ordered its hazardous site cleanup division to investigate.[43] Their testing found methane and arsenic in just one well, which "did not indicate levels of contaminants that would give EPA reason to take further action." Yet four independent scientists found elevated levels of methane and toxic chemicals related to hydrocarbons in local wells.

Today the town remains polarized over the subject of hydrofracking. Sampling of Dimock's water by universities and the EPA is ongoing.

In response to negative press, energy executives defended their process: "In sixty years of hydraulic fracturing across the country more than a million wells have been fracked," said Jim Smith, spokesman for the Independent Oil and Gas Association of New York. It has never, he added, "harmed a drop of water."[44]

The question of the source of methane remains an open one. As technology improves, it has become possible to identify the source of certain types of methane. Naturally occurring methane is considered "biogenic" (created by organic material decomposition), as opposed to "thermogenic" (created through the thermal decomposition of buried organic material). Biogenic methane is found at shallow depths, where water wells are typically drilled; energy companies usually pursue the deeper thermogenic methane. Through the use of isotope analysis, the methane in water can be identified as either biogenic or thermogenic, thereby determining if it is the result of natural causes or drilling.

A government study done in Colorado concluded that the methane gas tapped by drillers had migrated into dozens of water wells, possibly through natural faults and fissures exacerbated by hydrofracking. Dennis Coleman, an Illinois geologist and expert on molecular testing, has witnessed methane gas seeping underground for more than seven miles—many times what the industry says should be possible. "There is no such thing as 'impossible' in terms of migration," Coleman told ProPublica. "Like everything else in life, it comes down to the probability."[45]

Are the Chemicals in Hydrofracking Fluids Harmful?

The mantra of the energy industry is, as Energy in Depth puts it, that hydrofracking fluid is "greater than 99 percent...water and sand, and the fraction of what remains includes many common industrial and even household materials that millions of American consumers use every day."[46] Most of those chemicals, say industry boosters, are no more harmful than "what's underneath your kitchen sink."[47]

While some of the chemicals are common and benign—sodium chloride (used in table salt), borate salts (used in cosmetics), or guar gum (used to make ice cream)—others contain toxic additives—such as benzene (a carcinogen) or the solvent 2-Butoxyethanol, known as 2-BE.[48] While they comprise a tiny percentage of the mixture, hazardous exposure to some of these chemicals is measured in the parts per million.

The most common chemical, used in particular between 2005 and 2009, was methanol; other widely used chemicals included isopropyl alcohol, 2-Butoxyethanol, ethylene glycol, hydrochloric acid, petroleum distillates, and ethanol.[49] (For a list of chemicals known to have been used in hydrofracking, see the appendix at the end of this book.) Drillers tend to disclose only enough information about their fracking fluids to comply with worker-safety regulations. This usually consists of a product's trade name and rarely includes a complete list of constituents. A 2011 congressional report found that of 2,500 hydrofracking chemicals used, over 650 of them contained "known or possible human carcinogens, regulated under the Safe Drinking Water Act, or listed as hazardous air pollutants." The report also revealed that between 2005 and 2009, 279 products had at least one component listed as "proprietary" on their Occupational Safety and Health Administration (OSHA) material safety data sheet, meaning that the company that produced and used it chose not to make it public. The congressional committee noted that "Companies are injecting fluids containing unknown chemicals about which they may have limited understanding of the potential risks posed to human health and the environment."[50]

As for the small percentage of chemicals that are kept confidential, energy officials defend the use of trade secrets as necessary for innovation. This position is controversial, not least because every one million gallons of fluid blasted underground contains 10,000 gallons of chemicals.[51] Without knowing what chemicals are being used, it is impossible to test a site for them. While under the Safe Drinking Water Act the EPA regulated most types of underground fluid injection, the 2005 energy bill—permitting the "Halliburton Loophole"— stripped the agency of its authority to regulate hydraulic fracturing, and hence to determine whether the chemicals it uses are dangerous.

As noted in the previous chapter, the argument behind this special exemption was that state regulations sufficiently protect the environment, and that companies should be able to withhold the identity and amount of chemicals used as a trade secret. The result is that drilling regulation is left to a patchwork of state laws, and it is up to drillers to decide what constitutes a trade secret.[52]

This has hydrofracking opponents howling that the fox has been left to guard the chicken coop. According to the EPA website, "Several statutes may be leveraged to protect water quality, but EPA's central authority to protect drinking water is drawn from the Safe Drinking Water Act (SDWA). The protection of [drinking water] is focused in the Underground Injection Control (UIC) program, which regulates the subsurface emplacement of fluid."[53] The site goes on to point out that while the Energy Policy Act of 2005 "provided for exclusions to UIC authority," and specifically excluded hydrofracking from its regulation, some aspects of it, including the use of diesel fuel in fracking, was still regulated by the UIC program. In 1986, Congress enacted EPCRA—the Emergency Planning and Community Right to Know Act—a statute that requires drillers to maintain detailed information about each additive used in hydrofracking, information that should be available to federal, state, and local governments, to help first responders in case of an emergency (such as a well blowout). But fracking opponents say EPCRA does not help the average citizen identify potential pollutants, and the measure remains controversial—especially a caveat noted by the EPA: "All information submitted pursuant to EPCRA regulations is publicly accessible, unless protected by a trade secret claim."[54]

In studying hydrofracking fluids, Dr. Theo Colborn and her colleagues at the Endocrine Disruption Exchange in Colorado identified nearly 1,000 chemical products and some 650 individual chemicals in fracking fluids. At least 59 of these chemicals, and probably more, have been used to frack wells in New York State. Of these, 40 of the 59 chemicals (or 67.8 percent) had "the potential to cause multiple adverse health effects," and 19 (32.2 percent) cause "deleterious effects on the environment," according to a report by the American Academy of Pediatrics.[55]

How Is Flowback Disposed of?

As noted in chapter 1, about 33,000 natural gas wells are drilled each year, and 90 percent of them employ hydrofracking.[56] This translates into billions of gallons of potentially hazardous fluids being used annually. The chemicals and proppants added to the fluid, and naturally occurring contaminants, such as boron, barium, radium, and salts—including highly saline brine that dates to the Paleozoic era in the Marcellus Shale—stirred up by the drilling process.[57] (Salts can kill vegetation.) The result is flowback, a murky liquid, thick with salts, sulfur, chemicals, minerals, and proppants; it smells of sulfur, and sometimes contains low levels of radiation. Flowback is comprised of as little as 3 percent and as much as over 80 percent of the total amount of water and other materials used to fracture a well, according to the industry-backed website FracFocus. org.[58]

The flowback generally gets pumped into a pit, then into a separator tank that allows oil to surface; oil is skimmed off and sold. The remaining flowback needs to be carefully disposed of, which is where things get tricky.

Most states dispose of fluids by pumping them deep underground, into injection wells (which are distinct from gas and oil wells).[59] These deposits are regulated by EPA under the Safe Drinking Water Act. But the geologic formations under Pennsylvania—the East Coast's test case for hydrofracking—are unsuitable for injection wells.[60] Further, Pennsylvania produces so much wastewater that it threatens to overwhelm injection wells in neighboring Ohio.[61] Flowback and produced water are therefore commonly stored in large tanks, holding ponds (for evaporation), or, more often, sent by pipeline or truck to public wastewater treatment plants. This raises the chance of spills, traffic accidents, and wear and tear on roads, a major bone of contention for many rural communities. If the Marcellus and Utica Shales are opened to widespread hydrofracking, states like New York could produce hundreds of millions of gallons of flowback every day. That wastewater will likely be trucked to treatment plants. In Pennsylvania, that is already the case.

Yet most sewage treatment plants are not equipped to remove the chemicals, salts, Total Dissolved Solids (TDS), and radioactive elements in the drillers' wastewater. These contaminants can greatly increase the salinity of rivers and streams, which can harm aquatic life; affect the taste, smell, and color of tap water; interfere with the biological treatment process at sewage plants; and damage industrial and household equipment. Without a process to identify and test for these chemicals, it is impossible to know whether they are in drinking supplies.[62]

Pennsylvania promotes itself as hydrofracking-friendly, and former state DEP secretary John Hanger has said that "there are business pressures" on drillers to "cut corners....It's cheaper to dump wastewater than to treat it." [63] Yet Hanger insists that fears of water contamination from flowback are overblown. "Every single drop that is coming out of the tap in Pennsylvania today meets the safe drinking water standard," he maintains. But Hanger acknowledges that state water treatment plants are not equipped to treat flowback.[64]

In 2013, the federal EPA fined three Pennsylvania treatment plants for accepting hundreds of thousands of gallons of Marcellus Shale flowback that contained "multiple toxins and more than 7 million pounds of salt every month."[65] The plants discharged the water into the Allegheny River watershed, which provides drinking water to Pittsburgh and dozens of other communities. While the EPA penalty was only $83,000, the company that runs the treatment plants is temporarily banned from accepting further flowback and must invest $30 million to upgrade its facilities in order to comply with newly stringent state regulations.

Can Flowback Be Radioactive?

The simple answer is yes. The explosives and powerful pumps used by hydraulic fracturers exert enormous pressures, and sometimes loosen naturally occurring radioactive material called "radionuclides"—including radon, radium, thorium, and uranium—from subterranean rock. The flowback dislodges these elements from shale and sucks them up to the surface. Their radioactivity is low but measurable. Another source of radiation is man-made radionuclides, which are sometimes used as "tracers," to help define the injection profile, the fractures, and the fluid flow created by hydrofracking.[66]

Naturally occurring radionuclides are a subject of increasing concern. Their half-lives are longer than those of man-made isotopes, and they linger in the environment longer. Long-term exposure to radiation can have adverse health effects; even small amounts of exposure to radionuclides can be harmful.[67] When radon (a carcinogenic gas) and its byproducts decay, they can waft into the air, lodge in lungs, and cause lung cancer. If a gas well's radon-laced flowback mixes with drinking water, it can cause cancer of the internal organs, especially stomach cancer.

Industry officials and EPA regulators have played down the health risks of fracking.[68] But pediatricians affiliated with the Preventive Medicine and Family Health Committee of the state of New York have called for a moratorium on hydrofracking until its impacts on health are better understood.[69]

In 2011, the *New York Times* uncovered what it termed "never-reported studies" by the EPA and a "confidential study by the drilling industry" concluding that flowback radionuclides cannot be completely diluted.[70] The newspaper has also reported that the Pennsylvania DEP has turned a blind eye to these concerns, requesting, rather than requiring, gas companies to treat their own flowback rather than sending it to public water treatment facilities, for example.

What Are Injection Wells?

As mentioned above, most states outside the Northeast dispose of flowback by pumping it deep underground into injection wells regulated by the EPA.

In Texas, wastewater injection wells are becoming a common phenomenon. Truckloads of flowback run 24 hours a day, seven days a week, at a rate of 30 to 40 per day in some small south Texas towns. The amount of wastewater disposed of in state wells has jumped from 46 million barrels in 2005 to nearly 3.5 billion barrels in 2011, according to state regulators.[71] The state has over 8,000 active disposal wells, which is reportedly far more than Ohio or Pennsylvania. Texas has an additional 25,000 wells that use waste fluids to hydrofrack for additional oil and gas.[72]

Complaints to the Railroad Commission of Texas, the state's oil and gas regulator, argue that wastewater has spilled from pumps, tanks, and storage ponds, killing trees and vegetation. In 2005, flowback escaped from a disposal well and contaminated an aquifer, the Pecos River Cenozoic Alluvium. Remediation is ongoing; the company that operated the well had its permit revoked and declared bankruptcy.[73] While such cases are the exception rather than the rule, they undermine public trust in hydrofracking in general.

Does Hydrofracking Cause Earthquakes?

In addition to concerns over flowback and its potential for contamination of water sources, earthquakes remain a central focus point for opponents. The two are related through injection wells.

Hydrofracking has been blamed for causing earthquakes from Arkansas to England, though most of these tremors have been so minor as to be nearly undetectable. In fact, according to one British report, they "cause no more seismic activity than jumping off a ladder."[74] But that's not the entire picture. As the number of wastewater injection wells has risen since 2001, the number of earthquakes measuring 3.0 or higher on the Richter scale in midcontinent regions that are usually seismically quiet has surged—from 50 in 2009 to 87 in 2010 and 134 in 2011, representing a sixfold increase over last century—according to a US Geological Survey (USGS) report.[75]

Researchers from the Energy Institute at Durham University in England analyzed 198 reports of induced seismicity (minor quakes that are caused by human activity) since 1928. They found only three earthquakes they argued were caused by hydrofracking, all in 2011: one near Blackpool, England, one in the Holt River Basin in Canada (magnitude 3.8 on the Richter scale), and one in Oklahoma (magnitude 5.7).[76] The Oklahoma quake was caused by the injection of millions of gallons of flowback into deep rock formations. The water pressure built up and weakened the rock; sited near a geologic fault, this apparently set off tremors that hit Prague, Oklahoma, destroying 14 homes, and damaging almost 200 other buildings. The tremors were felt across 17 states.[77]

Lately, injection wells have been linked to a string of tremors in Arkansas, Oklahoma, Ohio, Texas, and Colorado. According to the USGS, only a fraction of roughly 40,000 waste-fluid disposal wells have caused earthquakes large enough to be of concern.[78] While the magnitudes of these quakes is generally small, USGS reports that their frequency is increasing.

What Are "Fugitive Emissions"?

As we've seen, once a well has undergone hydraulic fracturing, natural gas released from shale flows naturally up the borehole to the surface, where most of it is captured and put to use. But some of this gas, most of which is methane, escapes into the atmosphere. These leaks are called "fugitive emissions."

Methane can escape accidentally, through broken pipes, valves, or other equipment (diesel- or natural gas-powered compressors, drill rigs, pumps). A 2009 report in *Pipeline and Gas Journal* notes that while old, cast-iron pipes make up only 3 percent of US gas mains, they are responsible for most of the leaks—32 percent of methane emissions—from the natural gas distribution system. In 2012, some 3,300 gas leaks were discovered in Boston alone, according to a report in *Environmental Pollution*.[79]

Methane is sometimes released on purpose—when it is vented or not fully burned during flaring—in the refining process, and when being piped into homes or businesses. According to one report, leaks in Russian pipelines account for 0.6 percent of the natural gas transported.[80]

Fugitive emissions of methane in particular are hard to measure, and researchers disagree on the numbers. The EPA has estimated methane leaks at a rate of 2.3 percent of total production. But a recent study in Colorado and Utah found leakage rates of 4 percent and 9 percent, respectively.[81] The new study—conducted by the National Oceanic and Atmospheric Association and the University of Colorado at Boulder—offers only a "snapshot" of a specific location on a specific day, but if the true figure is at the upper end of the scale, opponents argue, natural gas is actually not cleaner than other fuels.

This question challenges one of the central rationales for promoting natural gas—that it burns cleaner than other fossil fuels—and, not surprisingly, has led to a rancorous dispute.

The previously mentioned 2011 methane study issued by Robert W. Howarth and colleagues at Cornell concluded that "3.6 percent to 7.9 percent of the methane from shale-gas production escapes to the atmosphere in venting and leaks over the lifetime of a well."[82] This represents up to double the amount from conventional gas wells. If Howarth is correct, shale gas would be more polluting than oil or coal. But Howarth's study has been heavily criticized by the energy industry, and there are those who disagree with his methodology, including his Cornell colleague Lawrence Cathles, who called the study "seriously flawed."[83] In response, Howarth issued new data in 2012 that backed up his original findings, and noted: "Compared to coal, the [climate] footprint of shale gas is at least 20 percent greater and perhaps more than twice as great on the 20-year horizon and is comparable when compared over 100 years."[84]

Yet other studies conducted by the US Department of Energy and Carnegie Mellon University show that emissions from shale gas are much smaller than Howarth found. Research published by the EPA in 2012 put the figure at 2.2 percent, only a little more than conventional gas, and its 2013 inventory reduced the agency's estimate further, to 1.5 percent.[85]

Answering questions about the true impact of fugitive emissions is important for both sides of the debate. As the International Energy Agency (IEA) cautions, in its report "Golden Rules for a Golden Age of Gas": "Greenhouse-gas emissions must be minimized both at the point of production and throughout the entire natural gas supply chain. Improperly addressed, these concerns threaten to curb, if not halt, the development of unconventional resources."[86]

What Are the Consequences of Fugitive Emissions?

There have been cases where release of airborne substances tied to hydrofracking has coincided with reports of health problems among residents. People living near shale gas drilling pads complain anecdotally of headaches, diarrhea, nosebleeds, dizziness, blackouts, muscle spasms, and other problems, as the documentary *GasLand* made clear. But the evidence isn't always conclusive.

A 2011 study of air quality around natural gas sites in Fort Worth found no health threats. But the air in DISH, Texas (a town that renamed itself in a marketing deal with a satellite company) was found to have elevated levels of disulphides, benzene, xylenes, and naphthalene—harmful chemicals traced back to hydrofrackers' compressor stations.[87]

In Garfield County, Colorado, volatile organic compound emissions increased 30 percent between 2004 and 2006; during the same period there was a rash of health complaints from local residents, ranging from headaches and nausea to adrenal and pituitary tumors.[88] But there are few, if any, epidemiological studies that might show that hydrofracking caused these problems.

New technology and regulations could cut methane leakage to less than 1 percent of total production and ensure that natural gas impacts the climate less than coal or diesel fuel, according to the World Resources Institute.[89] As of this writing, the EPA is finalizing new standards to control emissions from upstream oil and gas operations that would reduce fugitive methane by 13 percent in 2015 and 25 percent in 2035.[90]

Why Are Reports about Health Impacts Mostly Anecdotal?

In places such as Silt, Colorado, hydrofracking has allegedly caused serious health problems. In 2001, a gas well being fractured near the home of Laura Amos blew out, tainting her water supply with 2-BE. Amos developed a rare adrenal-gland tumor. State regulators fined EnCana Corporation, the operator, $99,400 because gas was found in Amos's drinking water. The company disputed this but did not fight it in court. Other state regulators said hydrofracking was not to blame, claiming that that the 2-BE that had poisoned Amos came from household cleaning products. In 2006, Amos accepted a multimillion-dollar settlement from EnCana, which also bought her property. As part of her settlement, she signed a nondisclosure agreement and has refused to discuss the case further.[91]

One reason for a lack of hard evidence about such cases is that once they have been settled the documents are sealed.[92] Industry groups—relying on a strict definition of fracking that does not include other aspects of the process, such as poorly drilled wells and leaking methane—steadfastly deny that hydrofracking pollutes air and water. This stance enrages opponents, who cite a litany of fines and penalties by regulators against frackers, and complain that the drillers' lack of transparency puts public and environmental health at risk.

Several research organizations and journalists have suggested that industry and governmental pressure have made reporting on hydrofracking difficult, and that environmental reports may have been censored. In 2011, the *New York Times* reported that the results of a 2004 EPA study may have been censored due to political pressure.[93] An early draft of that study had discussed the possibility of environmental threats due to hydrofracking, but the final report changed that conclusion.

Is Anyone Studying How Hydrofracking Impacts Animal and Human Health?

A 2012 study by a team of researchers at Cornell's College of Veterinary Medicine suggests that hydraulic fracturing has sickened or killed cows, horses, goats, llamas, chickens, dogs, cats, fish, and other animals.[94] The authors looked at 24 case studies in six shale-rich states and found that hundreds of cows in Colorado, Louisiana, New York, Ohio, Pennsylvania, and Texas died or gave birth to stillborn babies after being exposed to hydrofracking fluids. This is the first, and so far only, peer-reviewed report to suggest such a link.

The study noted that it was difficult to assess health impacts because of the Halliburton Loophole. The researchers recommended that all hydraulic fracturing fluids be disclosed; that animals—and their milk, cheese, eggs, and other products— near wells be tested; that water, soil, and air be monitored before and after drilling begins; and that nondisclosure agreements be limited.

Questions about fracking's impact on human health are even more controversial. A number of doctors and academics have expressed concern about potential long- and short-term health risks from gas production. While the EPA investigation of hydrofracking's impact on drinking water is a good start, they argue, broader studies on its effects on people are necessary. The idea is to bring academic rigor to the often emotional debate.

Researchers at Harvard are building a mapping tool to correlate gas-drilling operations with reports of nausea, headaches, and respiratory ailments. And a team of toxicologists from the University of Pennsylvania have organized researches from 17 institutions to review cases of sickness from people who live near drill pads, compressor stations, or wastewater ponds.[95]

The study will consider the toxicity of flowback wastewater; whether air quality is dangerously impacted by flaring gases; and whether industry's reliance on diesel fuel to power drills, compressors, and trucks creates an unhealthy environment. The first project surveyed Pennsylvania residents who live in the Marcellus Shale region about health symptoms. Future projects include assessing the health of people living in the Barnett Shale region of Texas, and an examination of how state gas-drilling laws impact public health issues.

Can the Gap between Hydrofracking's Proponents and Opponents Be Bridged?

Although energy companies continue to hydrofrack, and with impressive results, they have not won the public's confidence. At the heart of this lies the Halliburton Loophole, allowing drillers to keep certain fracking fluids secret. The International Energy Agency suggests that for oil and gas producers to make peace with their adversaries and move forward, they should take common-sense steps: improve transparency about the chemicals they use; engage communities better; monitor wells more effectively; toughen rules on well design and surface spills; manage water supplies carefully; and reduce methane emissions. The IEA reckons that implementing such measures would add just 7 percent to total well costs, and would go a long way toward pacifying critics.[96]

The hydrofracking industry is extremely competitive, and there appears to be dissent among drillers over how much to cooperate with one another and what steps to take to ameliorate public concern. Yet some in the industry seem inclined to heed the IEA's advice, and have advocated for greater disclosure. Chesapeake Energy, which once compared keeping the chemical makeup of its fracking fluids secret to "Coke protecting its syrup formula," now says that disclosure would promote meaningful dialogue. "We as an industry need to demystify," Chesapeake's then-CEO, Aubrey McClendon, acknowledged to an industry conference in 2012, "and be very upfront about what we are doing, disclose the chemicals that we are using, search for alternatives to some of the chemicals."[97]

In 2011, FracFocus.org—a national registry to which operators can voluntarily post details about the ingredients of their fracking fluids—was launched by the Interstate Oil and Gas Compact Commission (an association of states), and the Groundwater Protection Council (a group of state water officials), with backing from energy companies. Of 18 states that require drillers to disclose their chemicals, 11 require or allow them to be reported by FracFocus.[98] But the FracFocus database—which has over 35,000 records—was initially plagued by incorrect entries and was designed to search just one well at a time. Activists said that the definition of "trade secret" should be clarified, and that FracFocus should include historical information. In late 2012, FracFocus changed the way it collects data from companies, and said it intends to create an easier-to-use and more comprehensive database; the following year, the website made it possible for regulators to search and aggregate information for the first time.[99]

But disclosure remains a sticking point. In Texas, legislators passed the nation's first rules requiring public disclosure of fracking chemicals.[100] Designed to promote transparency, the rules were held up as a model for the rest of the country. Yet few drillers complied. Between April 2011 and early December 2012, Texas drillers used terms such as "secret," "confidential," or "proprietary" 10,120 times out of 12,410 hydraulic fractures reported to FracFocus, according to the *Austin Statesman*. In the Eagle Ford Shale, the major oil and gas play in south Texas, the trade secret exemption was used 2,297 times out of 3,100 hydraulic fractures. "I think it's a loophole big enough you can drive a frack truck through," observed Luke Metzger, director of Environment Texas. "If the companies argue that fracking is safe, why are they hiding behind these trade secret loopholes? If you're going to the doctor, you want to know what you might have been exposed to."[101]

In 2013 Ken Salazar, then secretary of the interior, reiterated that while the Obama administration is preparing to open public lands to fracking, "there has to be disclosure" of chemicals used in the process. "People need to know what's being injected into the underground. I tell people in the oil and gas industry that unless they embrace...disclosure, that it'll be the Achilles Heel of their industry."[102]

PART III

HYDROFRACKING

TODAY AND TOMORROW

7

THE FUTURE OF FRACKING

There is now a great deal of research and assessment being done on hydrofracking by government agencies, the industry, academics, inventors, and environmentalists, and the fruits of their labors will be revealed in the next few years.

While the politics and economics of fracking remain volatile, there appears to be a greater willingness to compromise on either side of the debate. In the meantime, scientists are gathering new data: the EPA will issue the preliminary results of its nationwide study in 2014, and impose new regulations limiting air pollution caused by hydrofracking in 2015. The industry has entered a new dynamic phase, and is maturing on almost every level.

How Much Gas and Oil Is There in American Shale Deposits?

The answer to this seemingly simple question is a moving target. In 2011 the EIA estimated that the nation had 827 trillion cubic feet of shale gas; a few months later, the agency sharply reduced the estimate to 482 trillion cubic feet. As of this writing in 2013, the EIA estimates the United States has roughly 665 trillion cubic feet of technically recoverable shale gas (fourth globally, behind China, Argentina, and Algeria).[1]

The US Geological Survey (USGS), meanwhile, revised its estimates of shale oil in the Bakken and Three Forks formations—which lie beneath Montana, North Dakota and South

Dakota—upward, to 7.4 billion barrels of undiscovered oil—double the previous estimate—and 6.7 trillion cubic feet of gas, triple the previous estimate. The EIA adds that the United States now has an estimated 48 billion barrels of shale oil reserves (second worldwide, after Russia's 58 billion barrels).

These revisions were made as information and technology changed, or previously unexplored formations, like Three Forks, gained attention. Moreover, the definition of what constitutes a "technically recoverable" resource is shifting.

Geologists typically gather data on how much gas a well produces at the outset of operations, and base their estimates on it. But those estimates can be too low or too high. Some estimates have included gas in pockets that are so deep or small, or in areas that are off-limits (i.e., critical watersheds or environmentally sensitive zones), that they are unlikely to be drilled. Energy analysts have criticized EIA and USGS estimates as overly optimistic. "If the country is going to embrace natural gas as the fuel of the future, there needs to be a lot more transparency in how these estimates are calculated and a more skeptical and informed discussion about the economics of shale gas," said Bill Powers, editor of the *Powers Energy Investor*.[2]

A group affiliated with the Colorado School of Mines doubts that the oft-quoted figure, that the United States has a "100 year supply" of natural gas, is realistic and says the nation may only have 23 years' worth of gas that is economically viable to recover.[3] And Gail Tverberg, an actuary who writes *Our Finite World*, a blog about resources, is skeptical about the hype around shale reserves: "The idea that the US is about to become a net oil exporter is simply a myth," she writes.[4]

According to a report in the *Wall Street Journal*, there is a growing consensus that the United States will see steady growth in natural gas supplies until they plateau sometime around 2040, after which they will slowly decline.[5] This conclusion was reinforced in February 2013, when an exhaustive study underwritten by the nonpartisan Alfred P. Sloan Foundation examined data from 15,000 wells in the Barnett Shale, near Fort Worth, Texas, plus research on shale in Pennsylvania, Louisiana, and Arkansas.[6] The study was among the first to examine both the geology and economics of hydrofracking, and it revealed that shale deposits can be highly variable in size and potential, even within the same region.

The gas industry uses the term Ultimate Economic Recovery (UER) to describe the economic performance of a well. Because no shale well has as yet undergone a complete life cycle, UER estimates are based on computer models and educated hunches.[7] While drillers are constantly on the hunt for new well sites, hydrofracking remains a risky business. The market can shift; productive wells can sit adjacent to unproductive wells; and some firms are contractually obliged to pump resources whether their wells turn a profit or not.

In the relatively large 5,000-square-mile Barnett play, for instance, energy companies have spent roughly $40 billion to lease and drill claims, yet critics such as petroleum geologist Art Beman say just a quarter of them are likely to prove economical.[8] Nonetheless, the Sloan study—which was praised by both pro- and anti-hydrofracking groups—suggests that in its lifetime the Barnett will likely see another 13,000 wells drilled, and produce some 44 trillion cubic feet of natural gas—equivalent to two years' worth of US natural gas consumption, and more than three times what has been produced there thus far.

How Political Has Hydrofracking Become?

Although hydraulic fracturing is highly partisan on the local level, it is a low-boiling issue on the national stage. In general, Republicans favor leaving most fracking regulation to the states, believing they can tailor rules to local needs, and do so more speedily than federal regulators. But many Democrats worry that without oversight cash-strapped states will engage in a "race to the bottom" and will trim their rules or fail to enforce them in order to compete for jobs and revenue.[9]

There is as yet no national standard for this industrial process, and 31 states with significant shale resources have responded with widely different requirements.[10] While a heavily fracked state like Pennsylvania requires full disclosure of chemicals, for instance, most states that are new to the process, such as Kansas, do not. In 2009, Ohio issued just a single hydrofracking permit. In 2010 it issued two. But in 2011 Ohio issued 42 permits, 27 of them between July and September. As a result, the Ohio Oil and Gas Association anticipates a $14 billion gain for the state by 2015.[11]

As of mid-2013, California had no rules specifically regulating fracking, and legislators were caught flat-footed. By that point, 851 wells had been hydrofracked in the state (mostly in Kern County, near Bakersfield); alarmed, California's strong environmental movement pushed hard for a moratorium. Now a dozen new laws are being debated by state legislators, and regulators are rushing to create a new set of regulations by 2014.[12]

In spite of such divisive politics, the hydrofracking debate has seen a few notable compromises. In 2013, Illinois was saddled with the worst pension problem in the nation—roughly $100 billion in unpaid pension liabilities—when a partisan fight erupted over hydrofracking.[13] Republicans, backed by industry, stumped for a loosening of regulations in order to attract natural gas producers. The Illinois Chamber Foundation said increased hydrofracking could bring more than 45,000 jobs to the state. But Democrats, backed by environmentalists, were deeply concerned about water depletion and pollution. In May, the state senate passed the nation's most stringent fracking regulations by a vote of 52 to 3. After "hundreds of thousands of hours" of negotiations, said Mike Frerichs, a Democrat from Champagne who sponsored the bill, the result was "tough regulations that are going to protect and preserve our most valuable resources.... We are going to increase home produced energy in our state in one of the most environmentally friendly ways possible."[14]

On the national stage, Democratic and Republican politicians alike take large campaign contributions from the gas and oil industry, and politicians of nearly every stripe seem to be pushing for the expansion of cheap gas. In his 2012 State of the Union address, President Obama said, "This country needs an all-out, all-of-the-above strategy that develops every available source of American energy—a strategy that's cleaner, cheaper, and full of new jobs. We have a supply of natural gas that can last America nearly one hundred years, and my Administration will take every possible action to safely develop this energy."[15]

The energy industry is quick to argue that the "shale gale" is the direct result of smart policy and intentional strategy. In a widely quoted op-ed piece in the *New York Times*, Christof Ruhl, BP's group chief economist, wrote that the American shale revolution "is not a happy accident of geology and lucky drilling." Rather, it comes from a particular set of circumstances that may be difficult to replicate elsewhere in the world. "The dramatic rise in shale-gas extraction and the tight-oil revolution," he wrote, "happened in the United States and Canada because open access, sound government policy, stable property rights and the incentive offered by market pricing unleashed the skills of good engineers." In Ruhl's view, policy and not geology is what is "driving the extraordinary turn of events that is boosting America's oil industry." While Asia, Latin America, and Africa have greater unconventional reserves than the United States, he noted, "the competitive environment, government policy and available infrastructure mean that North America will dominate the production of shale gas and tight oil for some time to come."[16]

Both Democrats and Republicans take credit for this policy, and agree that environmental and health questions need to be addressed for hydrofracking to succeed over the long term.

How Is Hydrofracking Being Regulated?

As mentioned above, the development and production of oil and gas is regulated by a matrix—to some, a crazy quilt—of federal, state, and local laws. Most federal laws are administered by the EPA or the Department of the Interior (DOI), though development of federal lands is overseen by the Bureau of Land Management (BLM) and the US Forest Service (USFS).

In the case of California and Illinois, state legislatures are scrambling to adapt to new technology and the public's mood swings, and it isn't always pretty. In 2012 alone over 170 bills to regulate oil and gas drilling were introduced in 29 states; but only 14 of them became law, according to the National Conference of State Legislatures.[17] Some state laws are tough, perhaps burdening the drilling industry unnecessarily. Others are lenient, perhaps leaving much of the country subject to environmental dangers.

The hydrofracking industry, meanwhile, argues that their operations are becoming cleaner and safer every day. According to a *Wall Street Journal* analysis of Pennsylvania DEP records from 2008 to 2012, the rate of environmental violations in the Marcellus Shale has dropped steadily as the industry matures.[18] Increasingly, large, well-funded, experienced companies are snapping up medium- and small-sized companies that don't have the resources or depth of knowledge to implement effective safety regimes. The analysis found that major energy firms—such as ExxonMobil, Shell, and Chevron—were cited for infractions in 6.5 percent of inspections; midsize companies—with a stock market value of $2 billion to $50 billion—were cited in 14 percent of inspections; and small firms—with a stock market value of less than $2 billion—were cited in 17 percent of inspections.

Opponents say that another factor—less aggressive regulation by Pennsylvania's pro-hydrofracking governor, Tom Corbett—could explain the drop in violations. Environmentalists point to a March 2011 memo that directed state DEP inspectors to clear all violation notices with senior department officials before issuing them, implying that the regulatory system had become politicized. The governor defended the practice and asserted that the decline in violations is the result of more rigorous inspections.[19]

In 2012 then-interior secretary Ken Salazar said that shale gas provided the United States the opportunity of energy independence, but added, "If we are going to develop natural gas from shale, it has to be done in a safe and responsible manner."[20] But when the DOI issued a new set of hotly anticipated rules governing hydrofracking on public lands in May 2013, environmentalists were dismayed. The new rules continue to allow energy companies to keep certain fracking chemicals secret, and allow them to run integrity tests on one representative well rather than all wells in a gas or oil field.

This ruling may be an indicator of how the Obama administration will regulate fracking going forward. The new rules were the first significant bit of regulation issued under the new interior secretary, Sally Jewell. (She worked in the oil industry in the 1970s, and is not afraid to say she fracked a few wells in Oklahoma.) She told reporters that it is critical for rules to keep up with technology, and that the federal government will continue to lease large tracts of public and Indian lands for energy development.[21]

What Steps Are Drillers Taking to Conserve Water?

"Water is now emerging as a significant opportunity and risk for oil and gas companies," said Laura Shenkar, an expert on corporate water strategy at the Artemis Project consulting firm.[22] In 2012, about 4.5 billion gallons of water were used for hydraulic fracking. By 2060, that number will spike to some 260 billion gallons, according to an estimate by Lux Research, a Boston consulting firm that monitors emerging technologies.[23]

As climate change and shifting weather patterns stress water supplies, a lack of water will impede hydrofracking. During the brutal drought of 2011, Texas regulators suspended water withdrawal permits in the Eagle Ford Shale, located near San Antonio.[24] In search of more efficient, sustainable water use, a few companies are experimenting with using recycled water. One company, Alpha Reclaim Technology, buys treated effluent from cities and towns in Texas and sells it to drillers in the Eagle Ford play.[25] Another plan is to develop mobile recycling units that will treat flowback, then reuse it. Companies in the same region are experimenting with using brackish water, a common underground resource in Texas. The drawback is that brackish water contains more salts and other elements, such as boron, which can harm the drilling process; some brackish reservoirs lie deep and are expensive to tap. A shallow well in the Eagle Ford play costs about $75,000 to drill, according to ConocoPhillips, while a deeper well could cost as much as $1 million.[26]

What Are "Green Completions"?

In an effort to make hydrofracking more environmentally friendly, the EPA instituted new "green completion" (or "reduced emission completion") rules in 2012, designed to cut down on air pollution. In essence, the nearly 600-page set of rules requires hydrofrackers to capture natural gas at the wellhead rather than flaring it off or releasing it into the atmosphere.

Green completions are used in the week or so between the initial drilling of a well and the time the well goes into production, a period when pollutants—and valuable methane—billow out of the borehole into the atmosphere, "like popping the top on a soda can," as the Natural Resources Defense Council (NRDC) puts it. That escaping gas represents half a trillion cubic feet of wasted gas annually, NRDC estimates.[27]

Every well is different, and there is no one-size-fits-all green completion process. When a well is drilled, it produces a mix of water, sand, hydrocarbon liquids, and gas. The elements are separated in a cylindrical vessel that allows the pressure from the well to drop: the liquids and solids drop while the gas rises. When hydrofracking dry gas, like that from the Marcellus Shale in New York, a green completion involves a two-step process to separate gas from flowback. In the case of wet gas, like that found in the Marcellus Shale in northern West Virginia, a three-phase separation separates gas from hydrocarbon liquids from flowback.[28]

Another target of green completion is the dramatic "flaring" of excess gas or oil. Typically used in a well's early production, this venting clears impurities from the well before production begins. But it also releases methane into the atmosphere. In composing the new rules, the EPA acted on its Clean Air Act mandate to reduce emissions of VOCs (volatile organic compounds) and potential carcinogens, such as benzene. The agency estimates that green completions would eliminate 95 percent of smog-forming VOCs emitted from over 13,000 new gas wells annually.[29]

During a green completion, wastewater is routed through a series of filters. A sand trap collects the solid materials, which are sent to a landfill. The remaining wastewater is cleaned, treated, and stored for reuse in the next drilling operation. The natural gas is then channeled into a pipeline, captured, and sold. Indeed, a "co-benefit" of the new rules is that, with the help of portable equipment to process gas and condensate (a low-density combination of hydrocarbon liquids that are present as gaseous compounds in natural gas), some 1 million to 1.7 million tons of methane can be recovered each year—gas that was previously wasted.[30] The EPA calculates green completions will yield $11 million to $19 million in savings per year.[31]

The agency has allowed the industry to delay full implementation of the rules until 2015, to ensure a smooth transition. Critics maintain the rules still have worrisome loopholes: existing facilities, for instance, are allowed to release one ton of benzene per year from specified equipment. Nor are industry experts happy with them. In fact, both sides of the fracking debate have threatened to sue the EPA over the new rules.

How Has the Business of Hydrofracking Evolved?

While the gas rush has benefited many Americans, it has not always made drillers and their investors rich. By 2013, exploration companies had poked so many boreholes into the ground and sucked so much natural gas out of shale that they produced a glut, dropped prices to near record lows, and left the hydrofracking industry with a hangover. "We just killed more meat than we could drag back to the cave and eat," Texas investment banker Maynard Holt, of Tudor Pickering Holt, who advised on many gas deals, lamented to the *New York Times*. "Now we have a problem."[32]

As noted in chapter 1, shale gas cost $13.68 per million BTU (MBTU) at Henry Hub, the main pricing point for American natural gas, in 2008. But by 2012, gas prices had fallen over 60 percent, to less than $2 per MBTU. By May 2013, prices had perked up, to $4.04 per MBTU, and the EIA expects the price to rise slightly, to $4.10 per MBTU, in 2014.[33]

Despite this slow recovery, hydrofracking is still not economical for many drillers. Most need the price to rise over $4 to cover costs, and the "sweet spot"—the price at which producers can make money while consumers are not too pinched—is about $5 or $6 per MBTU, analysts say.[34]

Because of the plunge in prices, the credit ratings and stock price of companies that had expanded rapidly into hydrofracking—such as Chesapeake Energy, Devon Energy, and Southwestern Energy—took a pounding. Many drillers took rigs out of production or shifted them to other regions.

In late 2012, the number of drill rigs exploring for natural gas fell by 30 percent, to 658, according to the energy services company Baker Hughes.[35] The steepest declines have been in the Fayetteville Shale in central Arkansas and the Haynesville Shale in Louisiana and Texas.

Some firms have shut down existing wells and halted new investments; others began to shed assets and retrench to pay down debts. By early 2013, Chesapeake's stock price had sunk, a raft of governance issues had caused the Securities and Exchange Commission to investigate the company, the board was shaken up, and Aubrey McClendon was removed as chairman. In an effort to raise $14 billion quickly, the company took an emergency unsecured bridge loan of $4 billion at 8.5 percent interest.[36] And Chesapeake shed valuable assets—such as large oil and gas operations in west Texas to Chevron and Royal Dutch Shell—in great haste.

Smaller companies like Norse Energy Corporation—which leased drilling rights to 130,000 acres in New York State, but was laden by debt and stymied by regulatory delays—have filed for bankruptcy, and are attempting to regroup under Chapter 11 protection.[37]

But many firms couldn't stop hydrofracking their wells even if they wanted to. Saddled by complex financial deals and lease arrangements they struck during the boom years, they are contractually obliged to proceed at full speed.

With the help of such Wall Street firms as Goldman Sachs, Barclays, and Jefferies & Company, the 50 biggest oil and gas firms raised some $126 billion between 2006 and 2012 and spent it on acquiring land and equipment, hydrofracking, and building infrastructure. This was double the companies' capital spending as of 2005, according to Ernst & Young.[38]

In 2008, before it hit serious financial turbulence, Chesapeake Energy signed up for aggressive "cash and carry" deals to help finance its growth. Plains Exploration paid $1.7 billion for ownership of one-third of Chesapeake's ownership of drilling rights it controlled in the Haynesville Shale. Plains further committed $1.7 billion to underwrite Chesapeake's drilling costs, in return for a percentage of profits. But while Chesapeake spent an average of $7,100 an acre on the drilling sites it leased in the Haynesville, Plains paid Chesapeake the equivalent of $30,000 an acre, according to the *New York Times*.

The bankers who orchestrated the deal made some $23 million on it.[39]

Drilling firms like Chesapeake, Petrohawk, and Exco Resources had also signed "use it or lose it" leases with landowners, which required them to start drilling within three to five years and begin paying royalties to the owner, or forfeit their drilling rights. As gas prices dropped, they were forced to spend a lot more money producing gas than they could sell it for; the economics no longer made sense.

"Quit drilling," T. Boone Pickens, the Dallas oil and gas billionaire, warned his fellow board members at Exco. In the 1990s, Pickens lost control of his energy company, Mesa, when prices dipped and he couldn't support the debt load. But Exco had made a $650 million deal with BG Group, an English gas company, when times were good. When gas prices slipped, Exco was contractually bound to keep its 22 rigs hitting production targets in the Haynesville Shale. Pickens was miffed. "We are stupid to drill these wells," he bluntly told the *New York Times*.[40]

A third element that has kept prices from rising involves geology. As mentioned in chapter 1, shale deposits can be "dry"—meaning they produce only natural gas (mostly methane)—or "wet"—meaning they also produce liquid natural gases (LNGs), such as ethane, butane, and propane. LNGs are used to make plastics, to power industrial heaters, or to produce the flames in your home barbecue. LNG prices are linked to the price of crude oil, which—unlike gas—is set globally and is relatively high.[41] As natural gas prices dropped, drill rigs have been shifted from dry to wet shales, boosting supplies; but this has also led to a glut of LNGs, and their prices will eventually fall.

In retrospect, it should have been obvious that with so much money flowing into drilling and so many new wells being hydrofracked in such a short amount of time, the natural gas bubble would burst. But, as with the Internet and real-estate bubbles of recent years, the savvy players were publicly predicting that gas supplies will last "for a hundred years" while privately expressing doubts.

In August 2008, Aubrey McClendon, then still CEO of Chesapeake Energy, told analysts that he had tamed the once risky, wildcatting oil and gas business and turned it into a regular, boring "manufacturing business that requires four inputs...land, people, science and, of course, capital." But as internal documents revealed by a lawsuit show, just two months later McClendon e-mailed his executives: "What was a fair price 90 days ago for a lease is now overpriced by a factor of at least 2x given the dramatic worsening of the natural gas and financial markets."[42]

Global behemoths like ExxonMobil and Royal Dutch Shell can afford to play for the long term and have the resources to cut costs and develop efficiencies. Gas prices will eventually rise, but for the near term the industry is suffering the aftermath of its rapid growth spurt.

What Technical Innovations Are on the Horizon?

"What was true yesterday is no longer true today," notes Andrew Place, of EQT Corp, a gas exploration firm based in Pittsburgh. "Systems are evolving....Public concerns have pushed the engineers to come up with solutions."[43]

Where some see risk, others see opportunity. By now hydrofracking has become a large, well-funded business that is not likely to disappear soon. Given this reality, a growing cadre of entrepreneurs and idealistic academics have been drawn to the industry with the intention of "doing well by doing good" in the shale oil and gas fields.

Graduate students, entrepreneurs, venture capitalists, and energy companies large and small are betting that more efficient hydrofracking technologies—such as new ceramic proppants designed with the help of nanotechnology by a Rice University consortium—and better wastewater management, which has caught the attention of global firms like

Schlumberger and small firms like Ecologix, of Alpharetta, Georgia, will be a profitable, expanding business for the foreseeable future. "Hopefully, we'll mend the dispute between environmentalists and oil companies by answering the wish list of both," Ecologix CEO Eli Gruber explained to the *Wall Street Journal*.[44]

Indeed, as I discuss in chapter 6, one of the most contentious aspects of hydrofracking is contained within its name: the use of millions of gallons of water—at least 70 billion to 140 billion gallons of H_2O annually, according to the EPA—to frack 35,000 wells a year.[45] This is a serious issue in this day of global warming and population growth. As a Schlumberger representative told a conference recently, if one million new wells are fracked worldwide by 2035, then reducing drillers' use of fresh water "is no longer just an environmental issue—it has to be an issue of strategic importance."

Some innovative companies have seen the writing on the wall, and asked: what else can we hydrofrack with?

In hundreds of small gas projects across Canada, and in a few test wells in the United States, drillers have replaced the water in their fluids with liquefied propane in gel form.[46] This substitution has a triple benefit: it preserves water supplies, which is increasingly important in drought-prone places like Texas, California, and Alberta; it limits the chance of polluting surface water and groundwater with spilled chemicals; and it removes the chance of setting off earthquakes by injecting wastewater into fault zones.

Using propane gel is essentially the same process as hydrofracking, with an added twist. As with water, the gel is pumped deep underground at tremendous pressure, which creates fissures in shale and releases bubbles of natural gas. Like water, the gel carries sand or man-made proppants to hold the fractures ajar so that gas can escape. But unlike water, the gel is turned into vapor deep underground by heat and pressure, and then flows to the surface with the natural gas, where it is recaptured; sometimes the used gel is reused or resold. Also, unlike water, propane does not wash chemicals or naturally occurring salts and radioactive elements back to the surface.

Gasfrac, a Canadian company, used propane in place of water in over 700 wells in 2012 in the provinces of Alberta, British Columbia, and New Brunswick, and has drilled test wells in several states, including Texas, Colorado, and Pennsylvania.[47] Boosters of propane fracking have grand ambitions and imagine the day, as one put it, when "the oil and gas industry could even be a net producer of water rather than a net user."

But wider use of propane gel is hampered by high cost, limited data about its effectiveness (largely because intensely competitive drilling companies are loathe to share information about innovations), and the industry's resistance to change.

The remediation (cleaning) of wastewater is another hydrofrack-related business that has seen meteoric growth recently. Ecosphere Technologies, of Stuart, Florida, uses a process called "advanced oxidation." In the chemical-free treatment, ozone is used to eliminate the chemicals used for bacteria control and scale inhibition, and recycles 10 percent of the water, according to the company.[48]

Other companies use "tunkey solutions," which allow drillers to clean water on site and to authenticate the results with tests. WaterTectonics, for instance, is a rapidly growing firm that uses electric currents to bind together contaminants, allowing them to be cleaned from the water. The company, which has a global licensing agreement to clean hydrofracking fluids for Halliburton, tripled its staff and finances between 2009 and 2011, the company said.[49] While the drop in gas prices impacted WaterTectonics, "the opportunity in frack water treatment is a very large market that is predicted to grow at an accelerated rate over the next ten years," said TJ Mothersbaugh, the company's business development manager.

What Are "Tracers," and How Are They Changing?

One of the most promising new ideas for reducing water contamination, or at least lowering the "dread-to-risk ratio" that shadows hydrofracking, is to inject fluids with "tracers" in order to track where they flow deep underground. But the public is wary of the man-made radioactive or chemical tracers currently used (mentioned in chapter 6), and now universities—such as Rice University in Houston—and companies—such as Southwestern Energy—are working to develop new technologies. The new tracers are stable, nontoxic, noninvasive chemicals with a unique "signature" for long-term fluid identification.

The mitigation of hydrofracking pollution is even attracting young, idealistic entrepreneurs. One approach under development by BaseTrace, a new company founded by a group of Duke University graduate students, uses an inert DNA-based tracer. DNA can provide a near-infinite number of sequence variations, so unique tracers can be tailored to individual wells. The BaseTrace product is robust and can withstand high temperatures and pressures, the company says. Just a thimbleful of the tracer can be detected, even when mixed with millions of gallons of fluid. The company hopes to introduce its tracer by late 2013 and is testing ways to identify groundwater pollution over long distances. "We really hope to make a difference by providing answers where previously there were none," wrote Justine Chow, BaseTrace's CEO.[50]

How Are Citizens Harnessing Big Data to Track Fracking?

Responding to a perceived decline of government oversight, SkyTruth, a nonprofit environmental monitoring group in West Virginia, created a real-time alert system that uses remote sensing and digital mapping to track pollution events. Founder Jon Amos says that "more and more the burden is going to be on the public to keep an eye on what's happening in the environment."[51]

SkyTruth has released a database created from over 27,000 industry reports on the chemicals used in hydrofracking, and made it freely available to the public. Monitoring the impacts of drilling in the Marcellus play in West Virginia, Pennsylvania, and New York, SkyTruth's alert system sends an e-mail update when a new event occurs. Alerting property owners that a natural gas well is about to be drilled nearby allows them to test their well water before, during, and after the hydrofracking process. "Our hope," said Amos, "is that this information will promote discussion."

In another instance of citizen initiative, Jamie Serra, who works for the Pennsylvania state legislature, created Fracktrack.org, a site that coalesces massive amounts of information about gas development in one site. The notion is to rely on thousands of data points to enhance transparency and understanding, to eliminate bias, and to help citizens understand what's happening around them.

"After seeing how many people were looking for information that existed but wasn't made readily available by the government," Serra said, he decided to "help complete missing and inaccurate data sets that are poorly designed and aren't mandated.... The numbers are no longer worth arguing over when we have the ability to generate and verify millions of responses in real-time."[52]

There are also ways for communities and even individuals to identify gas leaks and map drilling activities. These are not substitutes for effective regulation, but they can be useful supplements. Some concerned citizens have banded together as "methane monitors," to search out and report gas leaks around hydrofracking sites. They could potentially be compensated for this investigative work with rewards worth thousands of dollars under the Clean Air Act, much like "watershed watchdog" citizen groups help enforce Clean Water Act regulations.[53]

Josh Fox, the director of *GasLand*, and other "fracktivists" advocate the use of infrared video cameras to show methane emissions from gas and oil facilities. This can be an empowering tool, though one drawback is that people who don't properly decipher infrared images confuse standard heat emissions with methane emissions. Some experienced environmentalists, such as Walter Hang, who compiles data maps at his company, Toxic Targeting, warn that well-intentioned but error-prone citizen initiatives can undermine more professional efforts.[54]

What Are Federal Regulators Doing to Improve the Environmental Safety of Fracking?

In 2010, Congress requested that the EPA study the extent of hydrofracking's impact on the environment. In 2011, for the first time, the EPA chose seven natural gas plays on which to conduct case studies.[55] Two spots—in the Haynesville Shale, in DeSoto Parish, Louisiana, and in the Marcellus Shale in Washington County, Pennsylvania—were chosen because they had not yet been hydrofracked. The study also includes five places that have been hydrofracked—the Bakken Shale in Kildeer and Dunn counties, North Dakota; the Barnett Shale in Wise and Denton counties, Texas; the Marcellus Shale in Bradford and Susquehanna counties, Pennsylvania; the Marcellus Shale in Washington County, Pennsylvania; and Raton Basin in Las Animas County, Colorado. Results of the study are expected in 2014. "The value of these tests is that they are really the first independent review of what's happening from start to finish. It is a data set that doesn't really exist right now," Briana Mordick, a Natural Resources Defense Council scientist, has said.[56]

People on both sides of the debate agree that a broad and rigorous measurement of hydrofracking's impact on groundwater supplies is long overdue. Groundwater lies in deep aquifers far from sight, recharges slowly through precipitation from the surface, and can extend through subterranean chambers for hundreds of miles. (The Ogallala Aquifer, the nation's largest groundwater source, extends some 174,000 square miles beneath eight states. It has been in the news recently because the proposed Keystone XL pipeline was originally designed to cross over it in Nebraska, raising fears that an oil spill could contaminate the aquifer.) Once polluted, groundwater is notoriously difficult to clean. The crucial aspect to such a study would be a systematic sampling of the site prior to drilling, during drilling, and after drilling. This is accomplished with monitoring wells drilled in and around a hydrofracking site. The EPA is now in the midst of such a study.

To gauge the impact of hydrofracking on water supplies, the EPA is conducting a landmark nationwide study of 24,925 wells that were drilled with the process between September 2009 and October 2010. The agency opens its report with the assertion: "Natural gas plays a key role in our nation's clean energy future....However, as the use of hydraulic fracturing has increased, so have concerns about its potential human health and environmental impacts, especially for drinking water."[57] The study includes 18 research projects that will attempt to answer important questions about the use of water in five distinct stages of hydrofracking: from water acquisition to chemical mixing, well injection, flowback, and produced water, to wastewater treatment and waste disposal.

The study includes extensive mapping, a review of published literature, data analysis, scenario planning, computer modeling, laboratory studies, and case studies. It will focus on identifying ways that hydrofracking could contaminate drinking water on the surface and underground, including the role of elevated levels of methane. The study will not address other sensitive questions, such as possible links between hydrofracking and earthquakes or geochemical changes, however. (The Department of Energy and academic institutions are studying these questions.)

In January 2013, Chesapeake Energy agreed to let the EPA test an active drilling site (the EPA has not revealed the location of the test site). Range Resources has also agreed to let the EPA conduct tests at one of its wells in Washington County, Pennsylvania.[58] The companies may have calculated that if they can pass government inspection, public sentiment will shift in their favor, and government cooperation will become that much easier to win.

But the stakes are high: both regulators and industry executives say the EPA study will have a significant impact on the way hydrofracking is managed going forward.

Will Cars of the Future Run on Natural Gas?

A National Research Council report found that by increasing the efficiency of our vehicles, and using new technologies like biofuels and batteries, US cars and trucks could operate 50 percent more efficiently in 20 years.[59]

As we have seen, the shale gale has already changed the way the United States uses energy. Electric companies are forsaking coal for gas-powered turbines, and petrochemical companies are bringing their overseas facilities back to produce plastics in the United States for the first time in decades. But the holy grail of natural gas is in the gas tank of our cars.

Seventy percent of the oil consumed in America is used for transportation, a sector that emits more than 30 percent of our greenhouse gases.[60] Not only is natural gas cheaper than oil, but its emissions have a smaller impact than gasoline and diesel.

While many fleets of commercial trucks and city buses use natural gas, switching over passenger cars to natural gas is a much higher bar. There are four models of gas and dual-fuel cars for sale in the United States, and certified aftermarket conversion kits can be used on 40 models of cars and trucks, at a cost of $12,000 to $18,000.[61]

One drawback of natural gas cars is they cost thousands of dollars more than gasoline-powered ones—largely because the gas tanks must be bigger and heavier to store the fuel under pressure. (The Honda Civic GX, for instance, costs $5,200 more than a comparable gasoline vehicle.)[62] This could change in the future, however. Researches at companies like 3M are developing lighter fuel tanks or tanks that store natural gas at lower pressures.

There are few gas-powered cars available, which keeps costs high, and only 1,500 public fueling stations nationwide (only half of which are publicly accessible), according to the *Wall Street Journal*. But the economies of scale could build over time, especially once people understand how much money they will save. A comparable amount of natural gas can cost about half as much as gasoline, when it is at $4 a gallon (as of this writing in mid-2013, average gas prices nationwide are $3.78 per gallon, according to AAA).[63]

And there is the psychological hurdle. Many are afraid that natural gas will cause their cars to explode. This is not likely. For combustion, oxygen would need to mix with the methane in a gas tank and be ignited. CNG tanks are hardened against rupture and designed to vent rather than burst into flames.

In countries that lack gasoline-refining infrastructure— such as Pakistan and Iran—the governments have mandated a switch to natural gas.[64] In Russia—the world's second-largest gas producer after the United States—Gazprom, the state-owned energy monopoly, considers gas "a profitable core business" and is planning to create "a vast natural gas market" for cars, the company said in a statement.[65] Gazprom will be the exclusive supplier of natural gas for a new, green race-car series, the Volkswagen Scirocco R-Cup.

Some global energy firms, like Royal Dutch Shell, have turned natural gas into a low-sulfur diesel fuel that can be used in conventional cars and pumped at regular filling stations—though the process is not cheap.

With the right policy incentives in place, however, gas-powered vehicles could "increase the nation's energy security, decrease the susceptibility of the US economy to recessions caused by oil-price shocks, and reduce greenhouse-gas emissions," writes Christopher Knittel, a professor of energy and economics at MIT.[66]

Can Hydrofracking Help China, the World's Biggest Emitter of Greenhouse Gases, Reduce Its Carbon Footprint?

China generates about three-quarters of its electricity with coal-burning plants and produces twice as many greenhouse gases as the United States does every year. As the world's second-largest economy (after the United States), China increased its coal-fired generating capacity by 50 gigawatts in 2012, roughly equivalent to seven times the annual energy use in New York City. The country's breakneck growth is unlikely to slow down in the near term, and to keep pace it opens a new coal-powered plant each week. Consequently, the rate of China's greenhouse gas emissions increases 8 to 10 percent per year; by 2020 it will emit greenhouse gases at four times the rate of the United States.[67]

Climate scientists, such as those at the nonprofit research group Berkeley Earth, are alarmed by the environmental impact of China's growth and advocate that the United States help China switch from coal to natural gas.[68]

As noted, modern gas-fired power plants emit a third to a half of the carbon dioxide produced by coal plants producing the same amount of energy. China has vast shale formations and a budding gas industry. The EIA calculates that China has 1.3 quadrillion cubic feet of technically recoverable gas reserves in 2011, nearly 50 percent more than the United States has.[69] Yet China has limited knowledge of hydraulic fracturing, a voracious appetite for power, endemic corruption, and some of the worst pollution in the world. The government has recently begun to auction off drilling rights to shale gas plays in China, and many of the purchasers have little or no experience in energy production. Hydrofracking opponents fear the worst from this combination.

Yet, as most American drillers have shown, hydrofracking can be done in a relatively clean, responsible way. If China can set tough but fair environmental standards and enforce them, it will avoid delivering an unprecedented load of heat-trapping gases to an already overheated climate. Should China switch from coal to natural gas power, it could reduce its emissions by more than 50 percent, Berkeley Earth estimates.[70] It would also buy experts around the world time to develop new, cleaner, sustainable energy sources.

As the veteran environmental reporter Andrew Revkin has blogged for the *New York Times*, "This is how the world works, for better and worse." Revkin then offers his summary: "Energy needs and economic forces drive innovation, both in using energy more thriftily and finding new sources. Environmental issues arise. Pressure builds for change. Regulations and rules evolve. Industry resists. Lawsuits and environmental campaigns ensue. Innovations occur. And the human enterprise, often in herky-jerky fashion, moves forward."[71]

CONCLUSION

BEYOND HYDROFRACKING

The shale gas industry is still in its adolescence. And as adolescents are wont to do, it presents us with a dilemma: while the energy supplies in shale are too important to overlook, the potential health and environmental impacts of extracting them are too great to disregard. How we choose to respond to these contradictory priorities will have huge and long-lasting consequences.

The challenge is to learn how to produce the same amount of energy in a cleaner, safer way. As discussed, one of the main rationales in support of hydrofracking is that shale gas acts as a "bridge fuel" to ease the transition from dirty hydrocarbons to cleaner power supplies. The next energy revolution is likely to be based on clean, sustainable energy that will gradually supplant the Oil Age.

What Are "Renewables," and How Might They Affect Greenhouse Gas Emissions?

"Renewables"—energy from continually available supplies, such as sun, wind, moving water, and geothermal heat—are, in a sense, the earth's most basic energy source.

A 2011 estimate by the IEA claimed that most of the world's electricity could be provided by solar power within 50 years.[1] Indeed, many nations that once relied on a fuel mix akin to America's have made tremendous strides to replace fossil fuels with cleaner energy over the last two decades. While the United Sates generated 12.3 percent of its power with renewables in 2011, 13 other nations got at least 30 percent of their power from renewables, according to the IEA, and are aiming for even better results.[2] Iceland generated 100 percent of its electricity from hydroelectric and geothermal supplies; Norway got 97 percent of its power from hydro plants; Canada got 63.4 percent of its electricity from hydro and wind power; Portugal got 47 percent from renewables; and Spain got 30 percent from renewables.[3]

To be sure, many of these countries don't have the carbon-based energy resources that the United States does, so their citizens are accustomed to high energy costs. Some nations, like Portugal and Denmark, created financial incentives to drive the development of solar and wind power; others, like Germany, have strong green movements that made it politically palatable to keep energy prices high in order to reduce greenhouse gases.

But reducing carbon is as much about politics as it is about economics and engineering. Germany, which got 20.7 percent of its power from renewables in 2011, has a new set of incentives to push that share up to 35 percent by 2020.[4] But for the plan to work, Germany will have to bolster its electrical grid in order to transfer power generated in the windy north to the industrialized south more efficiently. That is expensive, and it comes in the midst of a global recession. So German leaders are carefully assessing public support for clean energy before committing to the new incentives.

In the United States, by contrast, reliance on fossil fuels seems an article of faith. Thanks to hydrofracking, cheap natural gas has made it easy to cling to that view. But as renewables become more commonplace and prices decline, this is an opportune moment to ask the kinds of questions raised in Europe.

"It's absolutely not true that we need natural gas, coal or oil—it's a myth," said Mark Z. Jacobson, an engineering professor at Stanford.[5] He and his colleagues have designed a renewable energy blueprint, which envisions New York—which is hardly as sun-baked as Nevada or as wind-swept as South Dakota—powered entirely by solar, wind, and hydro power by 2030. The blueprint calls for an energy mix: 10 percent from land-based wind; 40 percent from offshore wind; 20 percent from solar plants; and 18 percent from solar panels; plus a smattering of hydroelectric and geothermal power.

"You could power America with renewables from a technical and economic standpoint," Jacobson said. "The biggest obstacles are social and political—what you need is the will to do it."

A recent study by the National Renewable Energy Laboratory suggests that with targeted investments, emissions of CO_2 from US power plants could be reduced by as much as 80 percent by 2050.[6] In this scheme, most electricity would come from a combination of wind and solar, with gas-fired plants providing backup when the renewables are unable to meet peak demand (on a hot summer night, for example). Studies by researchers at Stanford University and other institutions have found that the United States has plentiful renewable resources; and, unlike Europe, the United States has enough open space to build sizable solar and wind generating plants.

Skeptics point out that renewables are not steady suppliers—the wind doesn't always blow and the sun doesn't always shine. They maintain that until a solution is found, we will rely on fossil fuels for the next several decades. Moreover, a shift in America from fossil fuels to wind, water, and sun would be expensive and cumbersome. To use renewable energy on a large scale requires modifying power grids and is generally more expensive than sticking with traditional carbon-based fuels.

Fatih Birol, chief economist at the IEA, says that as important as reducing greenhouse gases is, improving the energy efficiency of industry, vehicles, and homes is a quicker, easier approach than revamping the nation's grid.[7]

The transition to renewables will take time. It is likely that for quite awhile, users will rely on a combination of natural gas, other hydrocarbons, and renewable energy. The question remains, though: what will the ideal mix prove to be?

How Does the Low Price of Natural Gas Affect Renewables, and How Can Renewables Succeed?

The free market can instill a rigid discipline on energy markets, and one side-effect of fracking is that it provides so much cheap natural gas that it undermines the more expensive renewables it is supposed to provide a "bridge" to.

According to the Harvard researchers Michael McElroy and Xi Lu, the break-even price for electricity produced by a modern coal-fired plant is about 5.9 cents per kilowatt-hour. At $5 per million BTUs, the price of electricity from gas is about the same as that from coal; but when natural gas prices drop below $5 per MBTU, then coal cannot compete.[8]

The cost of producing wind power is about 8.0 cents per kilowatt-hour. Wind can compete with $5 per MBTU gas only if it can continue to benefit from the government's production tax credit (PTC), currently 2.2 cents per kilowatt-hour. If gas prices were to rise above $8.3 per MBTU, wind would be competitive even in the absence of the PTC. But in this case, coal plants would be cheaper than either gas or wind-generated power.

Thus, as McElroy and Lu pointed out in a *Harvard Magazine* article, the free market alone might not ensure that renewables succeed. If gas prices rise above $5 per MBTU, a carbon tax may be required to ensure that gas maintains its edge over coal. Similarly, should gas prices drop below $8.3 per MBTU, the PTC or similar initiatives might be required to support wind and solar energy.

If we are to achieve a low-carbon future, the Harvard researchers write, then gas prices must remain low enough to marginalize coal but not so low that renewable energy is rendered uncompetitive.

Fracking Is Here to Stay: How Will We Respond?

In a few short years, hydrofracking has fundamentally changed the energy landscape—for better or for worse. It has given us access to vast and previously inaccessible sources of natural gas and oil, provided jobs, stimulated the economy, lowered greenhouse gas emissions, altered the global marketplace, and changed policy. It is equally true that it adds methane to the air, pollutes water, dredges up toxic and radioactive substances, and has on occasion negatively impacted human and environmental health. As the hydrofracking boom widens, people are increasingly forced to weigh the benefits of shale resources against the costs of providing them.

To assess the pros and cons of hydrofracking, shale gas should be evaluated in the context of other readily available energy sources, especially coal. Coal is cheap and plentiful, but burning it emits twice the carbon dioxide per unit of energy as shale gas does and produces toxic metals, such as mercury, and other pollutants. Moreover, coal mining is more dangerous and environmentally destructive than hydrofracking for shale gas.

Based on what is known at this point, the shale gas delivered by hydrofracking is, on balance, better for the environment than coal. In the second decade of the twenty-first century, the United States is supplanting coal with natural gas. But Europe, China, and India are increasingly reliant on coal. Those nations have only just begun to investigate hydrofracking, and conditions there will make it more difficult to expand the use of shale gas as quickly as the United States is doing.

In America shale gas was seen as an overnight sensation, but in reality it took decades of research, testing, false starts, and tentative steps before it unlocked a new energy supply. For its part, Europe could take a decade or longer before it begins to use hydrofracking in earnest, if it ever does. There are many opportunities and pitfalls still to come.

Fracking is not inevitable. Important questions remain about the ultimate impact of the technology on health and the environment. Research is ongoing and should provide a useful guide in coming years. In the meantime existing techniques are being improved, and new technologies are being developed.

Hydraulic fracturing provides an epochal opportunity as a long-term supply of relatively clean fuel that will act as a bridge to more sustainable, renewable energy sources. But if it is used irresponsibly, undermines efforts to develop nonhydrocarbon energy, and does not supplant dirty fuels, then the great shale gale experiment will be judged a failure. With great opportunity comes great responsibility: how will we respond?

Appendix

LIST OF CHEMICALS USED IN HYDROFRACKING

Although there are hundreds of chemicals that can be used in the process of hydraulic fracturing as additives, what follows is a list of chemicals that are routinely used. This list was compiled by FracFocus.org.

Chemical name	CAS	Chemical purpose	Product function
Hydrochloric acid	007647-01-0	Helps dissolve minerals and initiate cracks in the rock	Acid
Glutaraldehyde	000111-30-8	Eliminates bacteria in the water that produces corrosive byproducts	Biocide
Quaternary ammonium chloride	012125-02-9	Eliminates bacteria in the water that produces corrosive byproducts	Biocide
Quaternary ammonium Chloride	061789-71-1	Eliminates bacteria in the water that produces corrosive byproducts	Biocide

(Continued)

Continued

Chemical name	CAS	Chemical purpose	Product function
Tetrakis hydroxymethyl-phosphonium sulfate	055566-30-8	Eliminates bacteria in the water that produces corrosive byproducts	Biocide
Ammonium persulfate	007727-54-0	Allows a delayed break down of the gel	Breaker
Sodium chloride	007647-14-5	Product stabilizer	Breaker
Magnesium peroxide	014452-57-4	Allows a delayed break down of the gel	Breaker
Magnesium oxide	001309-48-4	Allows a delayed break down of the gel	Breaker
Calcium chloride	010043-52-4	Product stabilizer	Breaker
Choline chloride	000067-48-1	Prevents clays from swelling or shifting	Clay Stabilizer
Tetramethyl ammonium chloride	000075-57-0	Prevents clays from swelling or shifting	Clay Stabilizer
Sodium chloride	007647-14-5	Prevents clays from swelling or shifting	Clay Stabilizer
Isopropanol	000067-63-0	Product stabilizer and/or winterizing agent	Corrosion inhibitor
Methanol	000067-56-1	Product stabilizer and/or winterizing agent	Corrosion inhibitor
Formic acid	000064-18-6	Prevents the corrosion of the pipe	Corrosion inhibitor
Acetaldehyde	000075-07-0	Prevents the corrosion of the pipe	Corrosion inhibitor
Petroleum distillate	064741-85-1	Carrier fluid for borate or zirconate cross-linker	Cross-linker
Hydrotreated light petroleum distillate	064742-47-8	Carrier fluid for borate or zirconate cross-linker	Cross-linker
Potassium metaborate	013709-94-9	Maintains fluid viscosity as temperature increases	Cross-linker

(Continued)

Continued

Chemical name	CAS	Chemical purpose	Product function
Triethanolamine zirconate	101033-44-7	Maintains fluid viscosity as temperature increases	Cross-linker
Sodium tetraborate	001303-96-4	Maintains fluid viscosity as temperature increases	Cross-linker
Boric acid	001333-73-9	Maintains fluid viscosity as temperature increases	Cross-linker
Zirconium complex	113184-20-6	Maintains fluid viscosity as temperature increases	Cross-linker
Borate salts	N/A	Maintains fluid viscosity as temperature increases	Cross-linker
Ethylene glycol	000107-21-1	Product stabilizer and/or winterizing agent	Cross-linker
Methanol	000067-56-1	Product stabilizer and/or winterizing agent	Cross-linker
Polyacrylamide	009003-05-8	"Slicks" the water to minimize friction	Friction reducer
Petroleum distillate	064741-85-1	Carrier fluid for polyacrylamide friction reducer	Friction reducer
Hydrotreated light petroleum distillate	064742-47-8	Carrier fluid for polyacrylamide friction reducer	Friction reducer
Methanol	000067-56-1	Product stabilizer and/or winterizing agent	Friction reducer
Ethylene glycol	000107-21-1	Product stabilizer and/or winterizing agent	Friction reducer

(Continued)

Continued

Chemical name	CAS	Chemical purpose	Product function
Guar gum	009000-30-0	Thickens the water in order to suspend the sand	Gelling agent
Petroleum distillate	064741-85-1	Carrier fluid for guar gum in liquid gels	Gelling agent
Hydrotreated light pretroleum distillate	064742-47-8	Carrier fluid for guar gum in liquid gels	Gelling agent
Methanol	000067-56-1	Product stabilizer and/or winterizing agent	Gelling agent
Polysaccharide blend	068130-15-4	Thickens the water in order to suspend the sand	Gelling agent
Ethylene glycol	000107-21-1	Product stabilizer and/or winterizing agent	Gelling agent
Citric acid	000077-92-9	Prevents precipitation of metal oxides	Iron control
Acetic acid	000064-19-7	Prevents precipitation of metal oxides	Iron control
Thioglycolic acid	000068-11-1	Prevents precipitation of metal oxides	Iron control
Sodium erythorbate	006381-77-7	Prevents precipitation of metal oxides	Iron control
Lauryl sulfate	000151-21-3	Used to prevent the formation of emulsions in the fracture fluid	Non-emulsifier
Isopropanol	000067-63-0	Product stabilizer and/or winterizing agent	Non-emulsifier
Ethylene glycol	000107-21-1	Product stabilizer and/or winterizing agent	Non-emulsifier
Sodium hydroxide	001310-73-2	Adjusts the pH of fluid to maintain the effectiveness of other components, such as cross-linkers	pH adjusting agent

(Continued)

Continued

Chemical name	CAS	Chemical purpose	Product function
Potassium hydroxide	001310-58-3	Adjusts the pH of fluid to maintain the effectiveness of other components, such as cross-linkers	pH adjusting agent
Acetic acid	000064-19-7	Adjusts the pH of fluid to maintain the effectiveness of other components, such as cross-linkers	pH adjusting agent
Sodium carbonate	000497-19-8	Adjusts the pH of fluid to maintain the effectiveness of other components, such as cross-linkers	pH adjusting agent
Potassium carbonate	000584-08-7	Adjusts the pH of fluid to maintain the effectiveness of other components, such as cross-linkers	pH adjusting agent
Copolymer of acrylamide and sodium acrylate	025987-30-8	Prevents scale deposits in the pipe	Scale inhibitor
Sodium polycarboxylate	N/A	Prevents scale deposits in the pipe	Scale inhibitor
Phosphonic acid salt	N/A	Prevents scale deposits in the pipe	Scale inhibitor
Lauryl sulfate	000151-21-3	Used to increase the viscosity of the fracture fluid	Surfactant
Ethanol	000064-17-5	Product stabilizer and/or winterizing agent	Surfactant
Naphthalene	000091-20-3	Carrier fluid for the active surfactant ingredients	Surfactant

(Continued)

Continued

Chemical name	CAS	Chemical purpose	Product function
Methanol	000067-56-1	Product stabilizer and/or winterizing agent	Surfactant
Isopropyl alcohol	000067-63-0	Product stabilizer and/or winterizing agent	Surfactant
2-Butoxyethanol	000111-76-2	Product stabilizer	Surfactant

NOTES

Preface

1. B. Pierce, J. Coleman, A. Demas: USGS press release: "USGS Releases New Assessment of Gas Resources in the Marcellus Shale, Appalachian Basin." Aug 23, 2011. (http://www.usgs.gov/newsroom/article.asp?ID=2893). And, Geology.com: "Marcellus Shale—Appalachian Basin Natural Gas Play."
   New research results surprise everyone on the potential of this well-known Devonian black shale.
2. Mireya Navarro, The New York Times: "Signing Drilling Leases, and Now Having Regrets," Sept 22, 2011. And Pressconnects. com: "Lawyers, landowners in fracking mineral rights 'force majeure' battle." Sept 3, 2010.
3. Edith Honan, Reuters: "NYC's Bloomberg opposes gas drilling in watershed," Jan 25, 2010.
4. Siena Research Institute, Siena College, Loudonville, NY: "Opposition to Fracking Falls Slightly as the Issue Continues to Nearly Evenly Divide Voters"(www.siena.edu.sri) May 20, 2013.
5. Allysia Finley, The Wall Street Journal: "Cuomo's Fracking Decision," Feb 12, 2013.
6. Clara A. Smith, LegislativeGazette.com: "Gandhi, Ruffalo lead anti-fracking rally in Albany," Feb 8, 2013. And Mireya Navarro, The New York Times: "Ruffalo Embraces a Role Closer to Home," Dec 2, 2011.

7. Abrahm Lustgarten, ProPublica: NYC: Gas Drilling Will Raise the Cost of Water by 30 Percent, Dec 16, 2008. http://www.propublica.org/article/new-york-city-calls-gas-drilling -effects-crippling-1216

8. Joint Landowners Coalition of New York http://www.jlcny.org/ site/index.php

9. Reid Pillifant, Capitalnewyork.com, "Why fracking has Cuomo at a loss," Feb. 13, 2013.

10. Freeman Klopott, Bloomberg.com, "New York Assemby Approves Two-year Moratorium on Fracking," Mar 6, 2013.

Introduction

1. David Brooks, New York Times: "Shale gas revolution," Nov. 3, 2011. The term "new energy bonanza" comes from Ian Bremer and Kenneth A. Hersh, New York Times: "When America stops importing energy," May 22, 2013.

2. US Environmental Protection Agency: "Natural gas: Electricity from natural gas." http://www.epa.gov/cleanenergy/energy- and-you/affect/natural-gas.html.

3. Tenille Tracy, Wall Street Journal: "Shale-gas estimate rises," Jun 19, 2013.

4. Jad Mouwad, New York Times: "Natural gas now viewed as safer bet," Mar 21, 2011.

5. Peg Mackey, Reuters: "U.S. to overtake Saudi as top oil pro- ducer: IEA," Nov. 12, 2012.

6. US Energy Information Administration, FAQ: "How much natu- ral gas does the United States have and how long will it last?" http://www.eia.gov/tools/faqs/faq.cfm?id=58&t=8.

7. Rob Reuteman, CNBC.com: "The math behind the 100-year, natural-gas supply debate," Jun 20, 2012.

8. Valerie Wood, Pipeline and Gas Journal, "Natural Gas Price Picture May Change By Late 2012," Sept 2011 (http://www.

pipelineandgasjournal.com/natural-gas-price-picture-ma
y-change-late-2012?page=show).

9. Felicity Carus, breakingenergy.com: "The Saudi Arabia of gas," May 23, 2011. http://breakingenergy.com/2011/05/23/the-saudi-arabia-of-gas/.

10. Artists against Fracking: "Don't Frack My Mother." http://artists-againstfracking.com/dont-frack-my-mother/.

11. David Letterman, The Late Show: Jul 27, 2012. http://www.youtube.com/watch?v=XGEc8KBKO8I.

12. New York Times, op-ed: "The Halliburton Loophole," Nov 2, 2009.

Chapter 1

1. California Energy Commission, Energy Quest: "Fossil fuels." http://www.energyquest.ca.gov/story/chapter08.html.

2. US Energy Information Administration: AEO2013 Early Release Overview, Dec 5, 2012. http://www.eia.gov/forecasts/aeo/er/early_elecgen.cfm.

3. US Department of Energy: "Coal: Our most abundant fuel." http://www.fossil.energy.gov/education/energylessons/coal/gen_coal.html.

4. Union of Concerned Scientists: "Environmental impacts of coal power: Air pollution." http://www.ucsusa.org/clean_energy/coalvswind/c02c.html.

5. Charles C. Mann, The Atlantic, National Journal: "What if oil lasts forever?," Apr 25, 2013. http://www.nationaljournal.com/energy/what-if-oil-lasts-forever-20130425?mrefid=site_search&page=1.

6. The Economist: "Coal in the rich world: The mixed fortunes of a fuel," Jan 5, 2013.

7. Bill Chappell: "Coal may pass oil as world's no. 1 energy source by 2017, study says," NPR, Dec 18, 2012.

8. The Economist: "Coal in the rich world."

9. Ibid.

10. Michael Birnbaum, Washington Post: "Europe consuming more coal," Feb 7, 2013.

11. The Economist: "Coal in the rich world."

12. William Tucker, American Spectator: "Environmentalist-in-chief," Jan 30, 2012. And Silvio Marcacci, CleanTechnica: "Sweeping State of the Union speech creates conflicting energy goals," Jan 25, 2012.

13. Brian Hardwick, NationalGeographic.com: "Obama unveils climate change strategy: End of line for U.S. coal power?" Jun 25, 2013.

14. The Economist: "Coal in the rich world."

15. California Energy Commission, Energy Quest: "Oil or petroleum." http://www.energyquest.ca.gov/story/chapter08.html.

16. Ibid.

17. Ibid. And OPEC: "Crude oil." www.opec.org/opec_web/en/press_room/180.htm. And Kathleen Brooks, Forex.com: "Why oil is measured in barrels." http://uk.finance.yahoo.com/news/why-oil-is-measured-in-barrels.html.

18. BP: Statistical Review of World Energy, Jun 2012. http://www.bp.com/content/dam/bp/pdf/Statistical-Review-2012/statistical_review_of_world_energy_2012.pdf.

19. Paul Tullis, Bloomberg Businessweek: " 'Peak oil' is back, but this time it's a peak in demand," May 1, 2013.

20. BP: Statistical Review of World Energy.

21. Wendy Lyons Sunshine, EnergyAbout.com: "Crude Oil basics." http://energy.about.com/od/Oil/a/Crude-Oil-Basics.htm.

22. US Energy Information Administration: Oil (petroleum). http://www.eia.gov/KIDS/energy.cfm?page=oil_home-basics-k.cfm.

23. Michael A. Levi, Ian W. H. Parry, Anthony Perl, Daniel J. Weiss, and Toni Johnson, Council on Foreign Relations: "Reducing US oil consumption," Jun 11, 2010.

24. Energy Information Administration: FAQ, "How much gasoline does the United States consume?" http://www.eia.gov/tools/faqs/faq.cfm?id=23&t=10.

25. Encyclopedia Britannica: "Natural gas." http://www.britannica.com/EBchecked/topic/406163/natural-gas/50586/History-of-use. And NaturalGas.org: History. http://www.naturalgas.org/overview/history.asp.

26. Chesapeake Energy: "Natural gas terminology." http://www.chk.com/naturalgas/pages/terminology.aspx.

27. US Energy Information Administration: "Natural gas explained." http://www.eia.gov/energyexplained/index.cfm?page=natural_gas_home.

28. US Department of Energy: *Modern Shale Gas Development in the United States: A Primer*, prepared by Ground Water Protection Council and ALL Consulting, April 2009. http://www.netl.doe.gov/technologies/oil-gas/publications/EPreports/Shale_Gas_Primer_2009.pdf.

29. Kate Galbraith, Texas Tribune: "Rich with natural gas, state eyes more oversight," Mar 11, 2011.

30. US Energy Information Administration: "How natural gas is used." http://www.eia.gov/energyexplained/index.cfm?page=natural_gas_use.

31. US Energy Information Administration: "Natural gas year-in-review 2008." http://www.eia.gov/pub/oil_gas/natural_gas/feature_articles/2009/ngyir2008/ngyir2008.html#note2.

32. Clifford Krauss and Eric Lipton, New York Times: "After the boom in natural Gas," Oct 20, 2012.

33. The Economist: "Bonanza or bane: Natural-gas prices are sure to rise—eventually," Mar 2, 2013.

34. US Energy Information Administration: Annual Energy Outlook 2013, Market Trends—Natural Gas. http://www.eia.gov/forecasts/aeo/MT_naturalgas.cfm.

35. Brad Plumer, Washington Post: "Is fracking a 'bridge' to a clean-energy future? Ernest Moniz thinks so." Mar 4, 2013. And John Podesta and Timothy E. Wirth, Center for American Progress: "Natural gas: A bridge fuel for the 21st century," Aug 10, 2009.

36. T. Boone Pickens: http://www.pickensplan.com/theplan.

37. Joe Romm, ThinkProgress.org: "Natural gas is a bridge to nowhere—absent a serious price for global warming pollution," Jan 24, 2012. http://thinkprogress.org/climate/2012/01/24/407765/natural-gas-is-a-bridge-to-nowhere-price-for-global-warming-pollution/?mobile=nc.

38. Mason Inman, National Geographic News: "Natural gas a weak weapon against climate change, new study asserts," Mar 14, 2012.

39. Winchester Action on Climate Change: "Unconventional hydrocarbons explained," http://www.winacc.org.uk/unconventional-hydrocarbons-explained.

40. US Energy Information Administration: "Technically recoverable shale oil and shale gas resources: An assessment of 137 shale formations in 41 countries outside the United States," Jun 13, 2013. http://www.eia.gov/analysis/studies/worldshalegas/.

41. Ibid.

42. Tenille Tracy, Wall Street Journal: "Shale-Gas Estimate Rises," Jun 19, 2013.

43. US Department of Energy: *Modern Shale Gas Development in the United States.*

44. Abrahm Lustgarten, ProPublica: "Hydrofracked? One man's mystery leads to a backlash against natural gas drilling," Feb 25, 2011. http://www.propublica.org/article/hydrofracked-one-mans-mystery-leads-to-a-backlash-against-natural-gas-drill/single.

45. Paul Stevens, Chatham House: "The 'shale gas revolution': Developments and changes," Aug 2012. http://www.chathamhouse.org/sites/default/files/public/Research/

Energy,%20Environment%20and%20Development/bp0812_
stevens.pdf.

46. US Department of Energy: *Modern Shale Gas Development in the United States.*

47. Shell.com: "Understanding Tight and shale gas." http://www.
shell.us/aboutshell/shell-businesses/onshore/shale-tight.html.

48. Wikipedia: "Shale oil." http://en.wikipedia.org/wiki/Shale_oil.

49. Canadian Society for Unconventional Resources: "Understanding tight oil." http://www.csur.com/sites/default/files/
Understanding_TightOil_FINAL.pdf. And Scholastic: "Alternate
energy sources." http://teacher.scholastic.com/scholasticnews/
indepth/upfront/grolier/alternate_energy.htm.

50. Investopedia.com: "Oil sands." http://www.investopedia.com/
terms/o/oilsand.asp. And Wikipedia: "Oil sands." http://
en.wikipedia.org/wiki/Oil_sands.

51. Shell.com: "Coal gasification." http://www.shell.com/global/
future-energy/unlocking-resources/coal-gasification.html.

Chapter 2

1. Energy in Depth: "Just the facts." http://energyindepth.org/
just-the-facts/.

2. Greenbang.com: "How much energy does the world use?" Mar 7,
2012. http://www.greenbang.com/how-much-energy-does-the-
world-use_21568.html.

3. Daniel Yergin, CNBC.com: "The Globalization of energy demand"
Jun 3, 2013. And Margo Habiby, Bloomberg.com: "Yergin predicts
global energy demand to rise 30% to 40% in next 20 years," Sep
13, 2010.

4. National Academies: "What you need to know about energy."
http://www.nap.edu/reports/energy/supply.html.

5. Wall Street Journal: "China tops U.S. in energy use," Jul 18, 2010.

6. Mark Weisbrot, Guardian: "2016: When China overtakes the US,"
Apr 27, 2011.

7. List compiled from EID: Just the Facts http://energyindepth.org/ just-the-facts/. And Wikipedia: "Hydraulic fracturing." http:// en.wikipedia.org/wiki/Hydraulic_fracturing#Uses.

8. Carl T. Montgomery and Michael B. Smith, Journal of Petroleum Technology (JPT): "Hydraulic fracturing: History of an enduring technology." (http://www.spe.org/jpt/print/archives/ 2010/12/10Hydraulic.pdf. And Alex Tremblath, Jesse Jenkins, Ted Nordhaus, and Michael Shellenberger, Breakthrough Institute: "Where the shale gas revolution came from," May 2012.

9. Ibid. And Fareed Zakaria, Washington Post: "Natural gas, fueling an economic revolution," Mar 29, 2012.

10. Ibid.

11. Ibid.

12. The Economist: "America's bounty: Gas works," Jul. 14 2012.

13. Douglas Martin, New York Times: "George Mitchell, a pioneer in hydraulic fracturing, dies at 94," July 26, 2013. Daniel Yergin, Wall Street Journal: "Stepping on the gas," Apr 2, 2011.

14. Edward E. Cohen, Aubrey K. McClendon, and Paul Gallay, Wall Street Journal: "The battle over fracking," Mar 26, 2012.

15. Ibid.

16. US Energy Information Administration: "Review of emerging resources: U.S. shale gas and shale oil plays," July 2011. http:// www.eia.gov/analysis/studies/usshalegas/pdf/usshaleplays. pdf.

17. Forbes: "George Mitchell." http://www.forbes.com/profile/george-mitchell/.

Chapter 3

1. George V. Chilingar, John O. Robertson, and Sanjay Kumar: *Surface Operations in Petroleum Production*. Amsterdam: Elsevier Science, 1989.

2. American Exploration & Production Council (AXPC): "The real facts about fracture stimulation: The technology behind America's

new natural gas supplies." http://www.axpc.us/download/ issues_and_info/natural%20gas/nat_gas_14apr2010.pdf. And ProPublica: "What is hydraulic fracturing?" http://www.propublica.org/special/hydraulic-fracturing-national.

3. Michael B. McElroy and Xi Lu, Harvard Magazine: "Fracking's Future," Jan–Feb 2013.

4. PennState Marcellus Center for Outreach & Research: http://www.marcellus.psu.edu/resources/maps.php; and Exploreshale. org. And Penn State Public Broadcasting and the Colcom Foundation: http://exploreshale.org.

5. Sen. John Kerry, Popular Mechanics: "Is fracking safe? The top 10 controversial claims about natural gas drilling," May 2010.

6. Michael B. McElroy and Xi Lu: "Fracking's future."

7. AXPC: "The real facts about fracture stimulation."

8. Edwin Dobb, National Geographic: "Bakken Shale oil: The new oil landscape," Mar 2013.

9. Rigzone.com: "How does directional drilling work?" http://www.rigzone.com/training/insight.asp?insight_id=295&c_id=1. Lynn Helms, DMR Newsletter: "Horizontal drilling." https://www.dmr.nd.gov/ndgs/newsletter/NL0308/pdfs/Horizontal. pdf. Wikipedia: "Directional drilling." http://en.wikipedia.org/wiki/Directional_drilling. David Blackmon, Forbes: "Horizontal drilling: A technological marvel ignored," Jan 28, 2013. http://www.forbes.com/sites/davidblackmon/2013/01/28/horizontal-drilling-a-technological-marvel-ignored/.

10. Hobart King, Geology.com: "Directional and horizontal drilling in oil and gas wells: Methods used to increase production and hit targets that can not be reached with a vertical well." http://geology.com/articles/horizontal-drilling/.

11. University of Texas at Arlington Natural Gas Program: http://www.uta.edu/ucomm/naturalgas/FrequentlyAskedQuestions.

php. Bill Toland, Pittsburgh Post-Gazette: "Drilling in the city: Lessons from Texas, part II," Mar 7, 2011.

12. Holly Deese and Robbie Brown, New York Times: "University of Tennessee wins approval for hydraulic fracturing plan," Mar 15, 2013.

13. Hobart King: "Directional and Horizontal drilling in oil and gas wells." Kevin Fisher, The American Oil & Gas Reporter: "Trends take fracturing 'back to future,'" Aug 2012. http://www.aogr. com/index.php/magazine/frac-facts.

14. King. "Directional and horizontal drilling in oil and gas wells."

15. Hobart King, Geology.com: "Mineral rights: Basic information about mineral, surface, oil and gas rights." http://geology.com/ articles/mineral-rights.shtml.

16. Edwin Dobb: "Bakken Shale oil."

17. Environmental Working Group: "Who owns the West: Oil & gas leases." http://www.ewg.org/oil_and_gas/part2.php.

18. T. R. Miller, ranchandresorttv.com: "Colorado laws on mineral rights," Apr 15, 2010. http://www.ranchandresorttv.com/f/ Water%20and%20Mineral%20Rights.

19. Anthony Andrews et al., Congressional Research Service: "Unconventional gas shales: Development, technology, and policy issues," Oct 30, 2009.

20. FracFocus.org: "Chemical use in hydraulic fracturing." http:// fracfocus.org/water-protection/drilling-usage.

21. Sourcewatch.org: "Fracking." http://www.sourcewatch.org/ index.php/Fracking.

22. Chesapeake Energy: "Hydraulic fracturing facts." http://www. hydraulicfracturing.com/Water-Usage/Pages/Information.aspx.

23. US Environmental Protection Agency: "Study of the potential impacts of hydraulic fracturing on drinking water resources: Progress report," Dec 2012. http://www2.epa.gov/ sites/production/files/documents/hf-report20121214.pdf.

24. William Ellsworth, Jessica Robertson, and Christopher Hook, US Geological Survey: "Man-made earthquakes," Jul 12, 2013. http://www.usgs.gov/blogs/features/usgs_top_story/man-made-earthquakes/.

Chapter 4

1. Railroad Commission of Texas: "Barnett Shale." http://www.rrc.state.tx.us/barnettshale/index.php.

2. Wikipedia: "Barnett Shale." http://en.wikipedia.org/wiki/Barnett_Shale.

3. Marianne Lavelle, National Geographic News: "Forcing gas out of rock with water," Oct 17, 2010. http://news.nationalgeographic.com/news/2010/10/101022-energy-marcellus-shale-gas-science-technology-water/. Independent Oil and Gas Association of New York: "The facts about natural gas exploration in the Marcellus Shale." http://www.marcellusfacts.com/pdf/homegrownenergy.pdf.

4. Marilyn Alva, Investor's Business Daily: "Range Resources ties fortune to Marcellus Shale." http://news.investors.com/business-the-new-america/071913-664449-range-resources-and-cabot-oil-play-marcellus.htm.

5. Don Warlick, Warlick Energy: "7 shale plays currently driving US drilling and development," Aug 1, 2012. http://warlickenergy.com/articles/7-shale-plays-currently-driving-us-drilling-and-development/.

6. Ibid. US Energy Information Administration: "Review of emerging resources: U.S. shale gas and shale oil plays," Jul 2011. http://www.eia.gov/analysis/studies/usshalegas/. Ron Nickelson, Clover Global Solutions: "The seven major US shale plays." http://www.usasymposium.com/bakken/docs/Clover%20Global%20Solutions,LP%20-%20The%20Seven%20Major%20US%20Shale%20Plays.pdf.

7. Noelle Straub, New York Times, Greenwire: "BLM suspends some oil and gas lease sales to review warming impacts," Apr 9, 2010.

8. Sourcewatch.com: "Marcellus Shale." http://www.sourcewatch. org/index.php?title=Marcellus_Shale#New_Jersey_State_ Senate_Passes_Fracking_Ban.

9. EcoWatch.com: "Ohio governor halts four more fracking wastewater injection wells after yesterday's quake," Jan 1, 2012. http://ecowatch.com/2012/breaking-ohio-govenor-halts-four-m ore-fracking-wastewater-injection-wells-after-yesterdays-quake/.

10. CNN.com: "Vermont first state to ban fracking," May 17, 2012. http://www.cnn.com/2012/05/17/us/vermont-fracking.

11. Mark Jaffe, Denver Post: "Colorado joins in suit to knock down Longmont fracking ban," Jul 11, 2013. http://www.denver-post.com/breakingnews/ci_23643679/state-joins-suit-knock-d own-longmont-fracking-ban. Ben Wolfgang, Washington Times: "I drank fracking fluid, says Colorado Gov. John Hickenlooper," Feb 12, 2013.

12. Gary Wockner, EcoWatch.com: "Colorado's bully governor says he will sue Fort Collins to overturn fracking ban," Feb 27, 2013. http://ecowatch.com/2013/fort-collins-overturn-fracking-ban/.

13. Blake Clayton, CFR.org: "Could the North American shale boom happen elsewhere?," Mar 15, 2013. http://blogs.cfr. org/levi/2013/03/15/could-the-north-american-shale-b oom-happen-elsewhere/.

14. Pavel Molchanov and Alex Morris, Raymond James: "Who'll be the first outside North America to commercialize shale gas?," Apr 18, 2011. http://www.anga.us/media/content/F7CEF91B-C242-1 1F8-265AC0573BFC90D8/files/raymond%20james_034616%20 (2).pdf.

15. Kari Lundgren, Bloomberg.com: "Shale gas explorer says U.K. production may start in 2014," May 10, 2012.

16. Katarzyna Klimasinska, Bloomberg.com: "European fracking bans open market for U.S. gas exports," May 23, 2012.

17. Detlef Mader: *Hydraulic Proppant Fracturing and Gravel Packing.* Amsterdam: Elsevier Science: 1989. Schumpeter, The Economist: "Spooked by shale: The shale-gas revolution unnerves Russian state capitalism," Jun 29, 2013. Wall Street Journal: "Government of Mongolia and Genie Energy sign strategic oil shale development agreement," Apr 19, 2013.

18. Schumpeter: "Spooked by shale."

19. Ibid.

20. US Energy Information Administration: "Technically recoverable shale oil and shale gas resources: An assessment of 137 shale formations in 41 countries outside the United States," Jun 2013. http://www.eia.gov/analysis/studies/worldshalegas/. Catherine T. Yang, National Geographic News: "China drills into shale gas, targeting huge reserves amid challenges," Aug 8, 2012. Bloomberg View: "How shale gas can save China from itself," Jul 11, 2013. http://www.businessweek.com/articles/2013-07-11/ bloomberg-view-how-shale-gas-can-save-china-from-itself.

21. Andrew C. Revkin, New York Times, Dot Earth: "A look at the role of policy in America's shale oil and gas era," Feb 6, 2013. http:// dotearth.blogs.nytimes.com/2013/02/06/a-look-at-the-role-of-p olicy-in-americas-shale-oil-and-gas-era/.

22. Tara Patel, Bloomberg.com: "France vote outlaws 'fracking' shale for natural gas, oil extraction," Jul 1, 2011. http://www.bloomberg.com/news/2011-07-01/france-vote-outlaws-fracking-sh ale-for-natural-gas-oil-extraction.html.

23. Novinite.com: "Bulgaria's GERB lifts Chevron shale gas license," Jan 17, 2012. Matthew Brown, Businessweek.com: "Fracking is flopping overseas," May 3, 2012.

24. BBC News: "Blackpool shale gas drilling suspended after quake," May 31, 2011. http://www.bbc.co.uk/news/uk-england-lancashire-13599161.

25. BBC News: "North American firms quit shale gas fracking in Poland," May 8, 2013. http://www.bbc.co.uk/news/business-22459629.

26. Rt.com: "Germany may ban fracking over environmental concerns," Feb 18, 2013.

Chapter 5

1. Edwin Dobb, National Geographic, "Bakken Shale oil: The new oil landscape," Mar 2013.

2. Clifford Krauss and Eric Lipton, New York Times: "The energy rush: After the boom in natural gas," Oct 20, 2012.

3. Ibid.

4. The Economist, Focus: "Peak oil," Mar 5, 2013. http://www.economist.com/blogs/graphicdetail/2013/03/focus-0.

5. Steven Mufson, Washington Post: "The new boom: Shale gas fueling an American industrial revival," Nov 14, 2012. http://articles.washingtonpost.com/2012-11-14/business/35506130_1_natural-gas-shale-cf-industries.

6. American Petroleum Institute: "Shale energy: 10 Points everyone should know." http://www.api.org/~/media/Files/Policy/Hydraulic_Fracturing/Hydraulic-Fracturing-10-points.ashx.

7. The Economist: "Deep sigh of relief: The shale gas and oil bonanza is transforming America's energy outlook and boosting its economy," Mar 16, 2013.

8. Adam Davidson, New York Times Magazine: "Welcome to Saudi Albany?," Dec 11, 2012.

9. The White House: "Remarks by the President in State of the Union Address," Jan 24, 2012. http://www.whitehouse.gov/the-press-office/2012/01/24/remarks-president-state-union-address.

10. Douglas Holtz-Eakin, New York Daily News, Opinion: "N.Y., start hydrofracking: Jobs await, and we all need cleaner, homegrown energy," Jul 13, 2011.

11. Robert Bradley, Post-Journal: "New York's energy bounty," Oct 16, 2011. http://www.post-journal.com/page/content.detail/id/592707/New-York-s-Energy-Bounty.html?nav=5071.

12. US Energy Information Administration: "U.S. Crude oil, natural gas, and NG liquids proved reserves," Aug 1, 2012. http://www.eia.gov/naturalgas/crudeoilreserves/.

13. Gary J. Schmitt, Weekly Standard: "Strategic gas," Apr 22, 2013. http://www.aei.org/article/foreign-and-defense-policy/defense/strategic-gas/.

14. American Petroleum Institute: "Shale energy."

15. The Economist: "America's bounty: Gas works," Jul 14, 2012.

16. Ibid.

17. Ibid.

18. Ibid.

19. Mufson: "The new boom."

20. Ibid.

21. Ibid.

22. Ibid.

23. Davidson: "Welcome to Saudi Albany?"

24. Honeywell.com, press release: "Honeywell completes acquisition of majority stake in Thomas Russell Co.; expands offerings for natural gas processing," Oct 22, 2012.

25. The Economist: "Gas works."

26. Mufson: "The new boom."

27. Ibid.

28. Jim Snyder, Bloomberg.com: "Fracking seen robbing OPEC of gasoline pricing power," Dec 10, 2012. http://www.bloomberg.com/news/2012-12-10/fracking-seen-robbing-opec-of-gasoline-pricing-power.html.

29. Ibid.

30. Ibid.

31. James F. Smith, Belfer Center, The Geopolitics of Energy Project, press release: "New study by Harvard Kennedy School researcher forecasts sharp increase in world oil production capacity, and risk of price collapse," Jun 2012. http://belfercenter.ksg.harvard.edu/publication/22145/new_study_by_harvard_kennedy_school_researcher_forecasts_sharp_increase_in_world_oil_production_capacity_and_risk_of_price_collapse.html.

32. Alan Riley, New York Times, op-ed: "The shale revolution's shifting geopolitics," Dec 25, 2012.

33. Christopher R. Knittel, MIT Center for Energy and Environmental Policy Research: "Reducing petroleum consumption from transportation," 2012.

34. Jackie Calmes and John M. Broder, New York Times: "Obama sets goal of one-third cut in oil imports," Mar 30, 2011. Matthew L. Wald, New York Times: "Energy secretary optimistic on Obama's plan to reduce emissions," Jun 27, 2013.

35. NaturalGas.org: "Natural gas and the environment." http://www.naturalgas.org/environment/naturalgas.asp.

36. Stanley Reed, New York Times: "Shell makes big bet on boom in natural gas," May 1, 2013.

37. Riley: "The shale revolution's shifting geopolitics."

38. The Economist: "Gas works."

39. Ibid. and Tom Fowler, Wall Street Journal: "America, start your natural-gas engines," Jun 18, 2012. http://online.wsj.com/article/SB10001424052702304192704577406431047638416.html

40. Keith Barry, Wired.com: "Staten Island Ferry goes green with natural gas," Jan 4, 2013.

41. Reed: "Shell makes big bet on boom in natural gas."

42. The Economist: "Gas works."

43. Ernest J. Moniz et al., MIT Energy Initiative: "The future of natural gas," Jun 6, 2011. http://mitei.mit.edu/publications/reports-studies/future-natural-gas.

44. Scott McNally, Scientificamerican.com: "Guest post: Water contamination—fracking is not the problem," Jan 25, 2012.

45. Houston Chronicle blog, Texas on the Potomac: "Citing new evidence, Democrats push for hydraulic fracturing hearings," May 27, 2011.

46. Energy in Depth: "Just the facts." http://energyindepth.org/just-the-facts/.

47. Department of Energy, Office of Fossil Energy and National Energy Technology Laboratory: *Modern Shale Gas Development in the United States:* A Primer (Report), April 2009. http://www.netl.doe.gov/technologies/oil-gas/publications/EPreports/Shale_Gas_Primer_2009.pdf.

48. Jesse Jenkins, TheEnergyCollective.com: "Energy facts: How much water does fracking for shale gas consume?" Apr 6, 2013. http://theenergycollective.com/jessejenkins/205481/friday-energy-facts-how-much-water-does-fracking-shale-gas-consume.

49. ConocoPhilips: "Why natural gas?," Sep 2011. http://www.powerincooperation.com/EN/Documents/11-2801%20NatGas-WhyNaturalGas.pdf.

50. New York Times, editorial: "The Halliburton Loophole," Nov 2, 2009.

51. Energy in Depth: "Just the facts." http://energyindepth.org/just-the-facts/.

52. US Environmental Protection Agency: "Clean energy: Natural gas." http://www.epa.gov/cleanenergy/energy-and-you/affect/air-emissions.html.

53. Guy Chazan, Financial Times: "Shale gas boom helps slash US emissions," May 23, 2012. http://www.ft.com/cms/s/0/3aa19200-a4eb-11e1-b421-00144feabdc0.html#axzz2aTaFwUDh.

54. US Energy Information Administration: "Annual energy outlook 2013," Apr 15, 2013.

55. ANGA, Natural Gas: "Smart power, smart politics," Oct 18, 2012. http://anga.us/blog/2012/10/18/natural-gas-smart-power-sm art-politics.

56. Ibid.

57. US Energy Information Administation: "U.S. energy-related CO_2 emissions in early 2012 lowest since 1992," Aug 1, 2012. http://www.eia.gov/todayinenergy/detail.cfm?id=7350.

58. The Economist: "The mixed fortunes of a fuel," Jan 15, 2013. http://www.economist.com/news/briefing/21569037-why-worlds-m ost-harmful-fossil-fuel-being-burned-less-america-and-more-e urope.

59. Tom Zeller Jr., New York Times, Green Blog: "Methane losses stir debate on natural gas," Apr 12, 2011. http://green.blogs.nytimes. com/2011/04/12/fugitive-methane-stirs-debate-on-natural-gas/.

60. Kevin Begos, Associated Press: "EPA methane report further divides fracking camps," Apr 28, 2013. Tina Gerhardt, Huffington Post: "Obama's climate action plan: Natural gas, fracking, methane and Mexico," Jul 1, 2013. http://www.huffingtonpost.com/ tina-gerhardt/obamas-climate-action-plan2_b_3523002.html.

61. Ibid.

62. Ibid.

63. Ibid.

64. Associated Press: "EPA lowers gas leak effect, climate harm from fracking," Apr 29, 2013. http://azstarnet.com/news/ science/environment/epa-lowers-gas-leak-effect-clima te-harm-from-fracking/article_a81cd5e4-aba4-5178-b 760-3379df6d0ba4.html.

Chapter 6

1. Mark Schlosberg, Food & Water Watch: "Why it's time for a global frackdown," Sep 21, 2012. http://www.foodandwaterwatch.org/ blogs/why-its-time-for-a-global-frackdown.

2. International Energy Agency: "Golden rules for a golden age of gas," May 29, 2012. http://www.iea.org/media/weowebsite/2013/unconventionalgasforum/0_2_BIROL.pdf.

3. Jenkins, TheEnergyCollective.com: "Energy facts: How much water does fracking for shale gas consume?," Apr 6, 2013. http://theenergycollective.com/jessejenkins/205481/friday-energy-facts-how-much-water-does-fracking-shale-gas-consume.

4. Chesapeake Energy: "Hydraulic fracturing facts, water usage." http://www.hydraulicfracturing.com/Water-Usage/Pages/Information.aspx.

5. Heather Cooley and Kristina Donnelly, Pacific Institute: "Hydraulic fracturing and water resources: Separating the frack from the fiction," Jun 2012.

6. Bruce Finley, Denver Post: "Colorado farms planning for dry spell losing auction bids for water to fracking projects," Apr 1, 2012. http://www.denverpost.com/environment/ci_20299962/colorado-farms-planning-dry.

7. Kate Galbraith, Texas Tribune: "In Texas, water use for fracking stirs concerns," Mar 8, 2013. http://www.texastribune.org/2013/03/08/texas-water-use-fracking-stirs-concerns.

8. Ibid.

9. Ibid.

10. Faucon, Benoît, Wall Street Journal: "Shale-gas boom hits Eastern Europe," Sep 17, 2012.

11. Alex Prud'homme, Men's Journal: "Bob Bea, the master of disaster," Feb 2013. http://www.mensjournal.com/magazine/bob-bea-the-master-of-disaster.

12. David Biello, Scientific American: "Natural gas cracked out of shale deposits may mean the U.S. has a stable supply for a century—but at what cost to the environment and human health?," Mar 30, 2010.

13. Abrahm Lustgarten, ProPublica: "Natural gas drilling: What we don't know," Dec 31, 2009. http://www.propublica.org/article/natural-gas-drilling-what-we-dont-know-1231.

14. Ibid.
15. Amy Mall, Natural Resources Defense Council, Switchboard: NRDC Staff Blog: "Incidents where hydraulic fracturing is a suspected cause of drinking water contamination," Dec 19, 2011.
16. Abrahm Lustgarten: "Natural gas drilling."
17. Ibid.
18. Ibid.
19. Valerie J. Brown, Environmental Health Perspectives: "Industry issues: Putting the heat on gas," Feb 2007. http://www.ncbi.nlm. nih.gov/pmc/articles/PMC1817691/.
20. US Environmental Protection Agency, news release: "EPA releases draft findings of Pavillion, Wyoming ground water investigation for public comment and independent scientific review," Dec 8, 2011. http://yosemite.epa.gov/opa/admpress.nsf/0/ EF35BD26A80D6CE3852579600065C94E.
21. Abrahm Lustgarten, ProPublica, "EPA's abandoned Wyoming fracking study one retreat of many," July 3, 2013. http://www. propublica.org/article/epas-abandoned-wyoming-fracking-st udy-one-retreat-of-many.
22. Russell Gold, Wall Street Journal: "EPA ties fracking, pollution," Dec 9, 2011. http://online.wsj.com/article/SB10001424052970203 50130457708647237334623 2.html.
23. Bill McKibben, New York Review of Books: "Why not frack?" Mar 8, 2012. http://www.nybooks.com/articles/archives/2012/mar/ 08/why-not-frack/.
24. Russell Gold: "EPA ties fracking, pollution."
25. Abrahm Lustgarten: "EPA's abandoned Wyoming fracking study one retreat of many."
26. Ibid.
27. Ibid.
28. Ibid.
29. Ibid.

30. Tina Gerhardt, Huffington Post: "Obama's climate action plan: natural gas, fracking, methane and Mexico," Jul 1, 2013. http://www.huffingtonpost.com/tina-gerhardt/obamas-climate-action-plan2_b_3523002.html.

31. R. L. Miller, Climate Hawks: "Why Lisa Jackson is leaving the EPA: The Keystone XL pipeline," Dec 27, 2012. http://www.dailykos.com/story/2012/12/27/1174315/-Why-Lisa-Jackson-is-leaving-the-EPA-the-Keystone-XL-pipeline.

32. Ramit Plushnick-Masti, Associated Press: "EPA backed off Weatherford water contamination probe after gas drilling company protested," Jan 16, 2013. http://www.dallasnews.com/news/state/headlines/20130116-epa-backed-off-weatherford-water-contamination-probe-after-gas-drilling-company-protested.ece.

33. Abrahm Lustgarten: "EPA's abandoned Wyoming fracking study one retreat of many."

34. Ian Urbina, New York Times: "A tainted water well, and concern there may be more," Aug 3, 2011. http://www.nytimes.com/2011/08/04/us/04natgas.html.

35. Susan Phillips, StateImpact: "Dimock: A town divided," Mar 28, 2012.

36. Associated Press: "Tests: Pa. gas drilling town's water still fouled," Oct 15, 2011.

37. Mike DiPaola, Bloomberg.com: "Fracking's toll on pets, livestock chills farmers: Commentary," Feb. 8, 2012.

38. StateImpact: "Dimock, PA: 'Ground zero' in the fight over fracking," http://stateimpact.npr.org/pennsylvania/tag/dimock/.

39. Lee O. Fuller, Energy in Depth, letter to Josh Fox: "Recommendations for GasLand 2," Aug 21, 2012. http://energyindepth.org/wp-content/uploads/2013/05/EID-Letter-to-JF-082112.pdf.

40. Susan Phillips, StateImpact: "Flaming taps: Methane migration and the fracking debate," Dec 19, 2011. http://stateimpact.npr.

org/pennsylvania/2011/12/19/flaming-taps-methane-migrat ion-and-the-fracking-debate/.

41. Josh Fox: "Affirming *GasLand*: A de-debunking document in response to specious and misleading gas industry claims against the film," Jul 2010. http://1trickpony.cachefly.net/gas/pdf/ Affirming_Gasland_Sept_2010.pdf.

42. StateImpact: "Dimock, PA."

43. Ibid.

44. Ibid.

45. Abrahm Lustgarten: "Natural gas drilling."

46. Energy in depth: "Just the facts." http://energyindepth.org/ just-the-facts/.

47. Ronald Bailey, Reason.com: "The top 5 lies about fracking," Jul 5, 2013. http://reason.com/archives/2013/07/05/the-top-5-lies-ab out-fracking.

48. US Department of Energy: "Modern shale gas development in the United States," Apr 2009. http://www.netl.doe.gov/technolo gies/oilgas/publications/EPreports/Shale_Gas_Primer_2009. pdf.

49. US House of Representatives Committee on Energy and Commerce Minority Staff: "Chemicals used in hydraulic fracturing," April 2011. http://democrats.energycommerce.house.gov/sites/default/ files/documents/Hydraulic-Fracturing-Chemicals-2011-4-18.pdf.

50. Ian Urbina, New York Times: "Chemicals were injected into wells, report says," Apr 16, 2011.

51. Jennifer Hiller, MySanAntonio.com: "Exact mix of fracking fluids remain a mystery," Feb 2, 2013. http://www.mysanantonio.com/ news/energy/article/Fracking-formulas-still-secret-4246634. php.

52. Abrahm Lustgarten: "Natural gas drilling."

53. EPA.gov: "Regulation of hydraulic fracturing under the Safe Drinking Water Act." http://water.epa.gov/type/groundwater/uic/class2/hydraulicfracturing/wells_hydroreg.cfm.

54. US Environmental Protection Agency: "Emergency Planning and Community Right-To-Know Act (EPCRA)." http://www.epa.gov/agriculture/lcra.html.

55. American Academy of Pediatrics: "Update on hydrofracking: A report by the American Academy of Pediatrics." http://aapdistrictii.org/update-on-hydrofracking.

56. Abrahm Lustgarten, ProPublica, "Hydrofracked? One man's mystery leads to a backlash against natural gas drilling," Feb 25, 2011. http://www.propublica.org/article/hydrofracked-one-m ans-mystery-leads-to-a-backlash-against-natural-gas-drill/single.

57. ScienceDaily: "Analysis of Marcellus flowback finds high level of ancient brine," Dec 18, 2012. www.sciencedaily.com/releases/2012/12/121218203537.htm.

58. FracFocus: "Fracturing fluid management." http://fracfocus.org/hydraulic-fracturing-how-it-works/drilling-risks-safeguards.

59. Terrence Henry and Kate Galbraith, New York Times/Texas Tribune: "As fracking proliferates, so do wastewater wells," Mar 28, 2013. http://www.nytimes.com/2013/03/29/us/wastewater-disposal-wells-proliferate-along-with-fracking.html.

60. Eliza Griswold, New York Times Magazine: "The fracturing of Pennsylvania," Nov 17, 2011.

61. Bob Downing, Akron Journal Beacon online: "Pennsylvania drilling wastes might overwhelm Ohio injection wells," Jan 22, 2013. http://www.ohio.com/news/local/pennsylvania-drilling-wastes-mi ght-overwhelm-ohio-injection-wells-1.367102.

62. Ian Urbina, New York Times: "Regulation lax as gas wells' tainted water hits rivers," Feb 26, 2011.

63. Ibid.

64. Andy Sheehan, Pittsburgh.CBSLocal: "State official: Pa. water meets safe drinking standards," Jan 2, 2011. http://pittsburgh. cbslocal.com/2011/01/04/state-official-pa-water-meets-safe-drinking-standards. John Hanger, Morning Call: "Pa. is monitoring Marcellus Shale wastewater," Jan 5, 2011. http://articles. mcall.com/2011-01-05/opinion/mc-dep-waste-water-hanger-yv-0106-20110105_1_drinking-water-wastewater-water-supply.

65. Don Hopey, Pittsburgh Post-Gazette: "EPA agrees to settle violations at water treatment plants in Venango, Indiana counties," May 23, 2013.

66. David B. Caruso, NBC Philadelphia: "44,000 barrels of tainted water dumped into Neshaminy Creek," Jan 3, 2011. http://www.nbcphiladelphia.com/news/local/Pennsylvania-Allows-Fracking-Tainted-Water-Dumping-Gas-Drilling-112804034.html. John C. Reis: Environmental Control in Petroleum Engineering. Gulf Professional Publishers, 1976.

67. Bob Weinhold, Environmental Health Perspectives: "Unknown quantity: Regulating radionuclides in tap water." http://www. ncbi.nlm.nih.gov/pmc/articles/PMC3440123.

68. Ian Urbina, New York Times: "Pressure limits efforts to police drilling for gas," Mar 3, 2011. http://www.nytimes.com/2011/03/04/us/04gas.html.

69. Sheila Bushkin-Bedient, Geoffrey E. Moore, and the Preventive Medicine and Family Health Committee of the State of New York: "Update on hydrofracking." http://aapdistrictii.org/update-on-hydrofracking.

70. Ian Urbina: "Regulation lax as gas wells' tainted water hits rivers."

71. Henry and Kate Galbraith, New York Times/Texas Tribune: "As fracking proliferates, so do wastewater wells," Mar 28, 2013. http://www.nytimes.com/2013/03/29/us/wastewater-disposal-wells-proliferate-along-with-fracking.html.

72. Ibid.

73. Ibid.

74. Nick McDermott, Daily Mail: "Fracking causes as much seismic activity as 'jumping off a ladder': Controversial method for extracting gas is 'extremely unlikely to trigger an earthquake we would feel,'" Apr 9, 2013.

75. Terrence Henry, StateImpact: "More on the science linking fracking disposal wells to earthquakes," Jul 12, 2012. http://stateimpact.npr.org/texas/2012/07/12/more-on-the-science-linking-fracking-disposal-wells-to-earthquakes/.

76. Energy & Environment: "Fracking report says fugitive emissions, not earthquakes are the real risk," Apr 10, 2013. http://www.eaem.co.uk/news/fracking-report-says-fugitive-emissions-not-earthquakes-are-real-risk.

77. Katie Davis, Daily Mail: "Largest ever man-made earthquake? Experts link 5.6 Oklahoma quake which buckled a highway and destroyed 14 homes to oil drilling," Mar 26, 2013. http://www.dailymail.co.uk/news/article-2299736/Largest-man-earthquake-Experts-link-5-6-Oklahoma-quake-buckled-highway-destroyed-14-homes-oil-drilling.html.

78. David J. Hayes, Deputy Secretary, US Department of the Interior, blog post: "Is the recent increase in felt earthquakes in the central US natural or manmade?" Apr 11, 2012. http://www.doi.gov/news/doinews/Is-the-Recent-Increase-in-Felt-Earthquakes-in-the-Central-US-Natural-or-Manmade.cfm.

79. Andrew C. Revkin, New York Times, Dot Earth: "Mapping gas leaks from aging urban pipes," Nov 20, 2012.

80. Stefan Lechtenbohmer et al., International Journal of Greenhouse Gas Control: "Tapping the leakages: Methane losses, mitigation options and policy issues for Russian long distance gas transmission pipelines," Oct 2007. http://www.sciencedirect.com/science/article/pii/S1750583607000898.

81. Steve Hamburg, Ecowatch: "Measuring fugitive methane emissions from fracking," Jan 4, 2013. http://ecowatch.com/2013/fugitive-methane-emissions-fracking/.

82. Robert W. Howarth, Renee Santoro, and Anthony Ingraffea, Climate Change, DOI: "Methane and the greenhouse-gas footprint of natural gas from shale formations: A letter," Mar 13, 2011. http://www.sustainablefuture.cornell.edu/news/attachments/Howarth-EtAl-2011.pdf.

83. Anne Ju, Cornell Chronicle: "Researchers challenge study on hydrofracking's gas footprint," Jul 29, 2013.

84. Andrew C. Revkin, New York Times, Dot Earth: "On shale gas, warming and whiplash," Jan 6, 2012. http://dotearth.blogs.nytimes.com/2012/01/06/on-shale-gas-warming-and-whiplash/.

85. Jon Entine, Forbes: "*New York Times* reversal: Cornell University research undermines hysteria contention that shale gas is 'dirty,' " Mar 2, 2012. http://www.forbes.com/sites/jonentine/2012/03/02/new-york-times-reversal-cornell-university-research-un dermines-hysteria-contention-that- shale-gas-is-dirty/. Climate Central: "Limiting methane leaks critical to gas, climate benefits," May 22, 2013. http://www.climatecentral.org/news/limiting-methane-leaks-critical-to-gas-climate-benefits-16020. America's Natural Gas Alliance, "Howarth: A Credibility Gap." (http://anga.us/links-and-resources/howarth-a-credibility-gap#.Uff0BxY1ZFI)

86. IEA: "Golden rules for a golden age of gas," Nov 2012. http://www.worldenergyoutlook.org/media/weowebsite/2012/gold-enrules/weo2012_goldenrulesreport.pdf.

87. Charles W. Schmidt, Environmental Health Perspectives: "Blind rush? Shale gas boom proceeds amid human health questions," Aug 2011. http://ehp.niehs.nih.gov/119-a348/#r13.

88. Ibid.

89. James Bradbury et al., World Resources Institute working paper: "Clearing the air: Reducing upstream greenhouse gas emissions from U.S. natural gas systems," Apr. 2013. http://www.wri.org/publication/clearing-the-air.

90. Vinesh Gowrishankar, Natural Resources Defense Council blog: "Clearing the air, and making clear more is needed to control methane leakage from natural gas," Apr 3, 2013. http://switchboard.nrdc.org/blogs/vgowrishankar/clearing_ the_air_and_making_cl.html.

91. Mike Soraghan, New York Times, Greenwire, "Baffled about fracking? You're not alone," May 13, 2011. http://www.nytimes.com/gwire/2011/05/13/13greenwire-baffled-about-fracki ng-youre-not-alone-44383.html.

92. Jim Efstathiou Jr. and Mark Drajem, Bloomberg.com: "Drillers silence fracking claims with sealed settlements," Jun 6, 2013.

93. Ian Urbina, New York Times: "Pressure limits efforts to police drilling for gas," Mar 3, 2011. http://www.nytimes.com/2011/03/04/us/04gas.html.

94. Krishna Ramanujan, Cornell Chronicle: "Study suggests hydrofracking is killing farm animals, pets," Mar 7, 2012. http://www.news.cornell.edu/stories/2012/03/reproductive-problems-de ath-animals-exposed-fracking.

95. Jon Hurdle, New York Times, Green Blog, "Taking a harder look at fracking and health," Jan 21, 2013. http://green.blogs.nytimes.com/2013/01/21/taking-a-harder-look-at-fracking-and-health/.

96. The Economist: "Fracking great: The promised gas revolution can do the environment more good than harm," Jun 2, 2012. http://www.economist.com/node/21556249.

97. Abrahm Lustgarten: "Natural gas drilling."

98. Mark Drajem, Bloomberg.com: "FracFocus drillers' registry to create chemicals database," May 23, 2013. http://www.

bloomberg.com/news/2013-05-23/fracfocus-drillers-registry-to-create-chemicals-database.html.

99. Hiller: "Exact mix of fracking fluids remain a mystery."

100. Jennifer Hiller, Austin Statesman: "Frackers avoid fluid disclosure despite new law," Feb 9, 2013. http://www.statesman.com/news/news/opinion/frackers-avoid-fluid-disclosure-despite-new-law/nWHKZ/.

101. Ibid.

102. David R. Baker, SFGate blog: "Salazar: Frackers should disclose chemicals," Mar 13, 2013. http://blog.sfgate.com/energy/2013/03/13/salazar-frackers-should-disclose-chemicals/.

Chapter 7

1. US Energy Information Administration: "Technically recoverable shale oil and shale gas resources: An assessment of 137 shale formations in 41 countries outside the United States," Jun 10, 2013. http://www.eia.gov/analysis/studies/worldshalegas/#.UbXmXBnIKs8.twitter. Zach Colman, The Hill, "Energy agency: US oil-and-gas reserves up 35 percent, thanks to shale boom," Jun 10, 2013. http://thehill.com/blogs/e2-wire/e2-wire/304495-eia-shale-boom-drives-us-oil-and-gas-reserves-up-35-percent. Ian Urbina, New York Times: "New report by agency lowers estimates of natural gas in U.S.," Jan 28, 2012.

2. Ian Urbina, New York Times: "Geologists Sharply Cut Estimate of Shale Gas," Aug 24, 2011. http://www.nytimes.com/2011/08/25/us/25gas.html.

3. Jeff Goodell, Rolling Stone: "The big fracking bubble: The scam behind Aubrey McClendon's gas boom," Mar 1, 2012. http://www.rollingstone.com/politics/news/the-big-fracking-bubble-the-scam-behind-the-gas-boom-20120301.

4. Gail Tverberg, "Our finite world: The myth that the US will soon become an oil exporter," Apr 16, 2012. http://ourfiniteworld.

com/2012/04/16/the-myth-that-the-us-will-soon-become-an-oil-exporter/.

5. Russell Gold, Wall Street Journal: "Gas boom projected to grow for decades," Feb 27, 2013. http://online.wsj.com/article/SB1000142 4127887323293704578330700203397128.html.

6. Ibid.

7. Agathena, Daily Kos: "To the gas industry: 'What the frack are you doing to our air and water?,'" Aug 9, 2012. http://www.dailykos.com/story/2012/08/09/1116090/-To-the-Gas-Industry-What-the-frack-are-you-doing-to-our-air-and-water.

8. Russell Gold: "Gas boom projected to grow for decades."

9. Lisa Margonelli, Pacific Standard: "The debate we aren't having," Mar 5, 2013. http://www.psmag.com/environment/the-energy-debate-we-arent-having-53400/.

10. Brad Plummer, Washington Post: "How states are regulating fracking (in maps)," Jul 16, 2012. http://www.washingtonpost.com/blogs/wonkblog/wp/2012/07/16/how-states-are-regulating-fracking-in-maps/.

11. Karen Showalter, PriceofOil.org: "New EPA regs address fracking…in 2015." http://priceofoil.org/2012/04/20/new-epa-regs-address-fracking-in-2015/.

12. Source Watch: "California and fracking." http://ftp.sourcewatch.org/index.php?title=California_and_fracking. Aaron Sankin, Huffington Post: "California fracking moratorium: Trio of bills would halt controversial drilling practice," May 1, 2013. http://www.huffingtonpost.com/2013/05/01/california-fracking-moratorium_n_3194576.html. Norimitsu Onishi, New York Times: "Fracking tests ties between California 'oil and ag' interests," Jun 1, 2013. http://www.nytimes.com/2013/06/02/us/california-oil-and-ag-face-rift-on-fracking.html.

13. Crain's Chicago Business: "Illinois Republicans push for fracking bill vote," Apr 30, 2013. http://www.chicagobusiness.com/

article/20130430/NEWS02/130439971/illinois-republicans-p ush-for-fracking-bill-vote.

14. Kerry Lester, Associated Press: "Illinois fracking bill: State legislature passes nation's toughest fracking regulations," May 31, 2013.

15. The White House: "Remarks by the president in State of the Union Address," Jan 24, 2012. http://www.whitehouse.gov/ the-press-office/2012/01/24/remarks-president-state-un ion-address.

16. Christof Ruhl, New York Times, Opinion: "Spreading an energy revolution," Feb 5, 2013. http://www.nytimes.com/2013/02/06/ opinion/global/spreading-an-energy-revolution.html.

17. National Conference of State Legislatures: "States take the lead on regulating hydraulic fracturing: Overview of 2012 state legislation." http://www.ncsl.org/issues-research/energyhome/regulating-hydraulic-fracturing-legislation.aspx.

18. Daniel Gilbert and Russell Gold, Wall Street Journal: "As big drillers move in, safety goes up," April 2, 2013. http://online.wsj. com/article/SB10001424127887324582804578346741120261384. html.

19. Ibid.

20. Balazs Koranyi, Reuters: "U.S. needs federal fracking rules—Salazar," Jun 25, 2012. http://articles.chicagotribune.com/ 2012-06-25/news/sns-rt-us-energysalazar-interview-pix-20120625_1_fracking-natural-gas-shale-gas.

21. John M. Broder, New York Times: "New fracking rules proposed for U.S. land," May 16, 2013. http://www.nytimes.com/2013/05/17/ us/interior-proposes-new-rules-for-fracking-on-us-land.html.

22. Erica Gies, New York Times: "Race Is on to clean up hydraulic fracturing," Dec 4, 2012.

23. Ibid.

24. Ibid.

25. Kate Galbraith, Texas Tribune: "In Texas, water use for frack-
    ing stirs concerns," Mar 8, 2013. http://www.texastribune.
    org/2013/03/08/texas-water-use-fracking-stirs-concerns/.
26. Ibid.
27. David Donger, NRDC Switchboard blog: "Leading companies
    already meet EPA's 'fracking' air pollution standards," Apr 18,
    2012. http://switchboard.nrdc.org/blogs/ddoniger/leading_
    companies_already_meet.html.
28. Pam Kasey, State Journal: "Expert: Green completion technology
    varies by natural gas region," May 7, 2012. http://www.statejour-
    nal.com/story/18165012/expert-green-completion-technology-v
    aries-by-natural-gas-region.
29. Andrew Maykuth, Philadelphia Inquirer: "Natural gas produc-
    ers turn to 'green completion,'" Nov 26, 2012. http://articles.
    philly.com/2012-11-26/business/35348948_1_natural-gas-sh
    ale-gas-marcellus-shale.
30. Ibid.
31. EPA: "Overview of final amendments to air regulations for the
    oil and natural gas industry," Apr 17, 2012. http://www.epa.gov/
    airquality/oilandgas/pdfs/20120417fs.pdf.
32. Clifford Krauss and Eric Lipton, New York Times: "After
    the boom in natural gas," Oct 20, 2012. http://www.
    nytimes.com/2012/10/21/business/energy-environment/
    in-a-natural-gas-glut-big-winners-and-losers.html.
33. The Economist: "America's cheap gas: Bonanza or bane,"
    Mar 2, 2013. http://www.economist.com/news/finance-
    and-economics/21572815-natural-gas-prices-are-sure-
    riseeventually-bonanza-or-bane. Reuters: "EIA raises U.S.
    natgas production estimate for 2013," Jun 11, 2013. http://
    www.reuters.com/article/2013/06/11/eia-outlook-natgas-
    idUSL2N0EN12L20130611.

34. J. Lester, Cleantech Finance: "The DOE and the exporting of liquefied natural gas," Mar 6, 2013. http://www.cleantechfinance. net/2013/doe-lngexports.

35. Jonathan Fahey, Associated Press: "Natural gas glut means drilling boom must slow," Apr 8, 2012. http://usatoday30.usatoday.com/ MONEY/usaedition/2012-04-09-APUSNaturalGasGlut_ST_U. htm.

36. John Kemp, Reuters: "McClendon's exit will not solve Chesapeake's problems: Kemp," Jan 31, 2013 http://www.reuters. com/article/2013/01/31/us-column-kemp-chesapeake-us gas-idUSBRE90U0K720130131.

37. Marcellus Drilling News: "New rough patch for Norse Energy: Force majeure lawsuit." http://marcellusdrilling.com/ 2013/05/new-rough-patch-for-norse-energy-force-majeure-lawsuit/.

38. Clifford Krauss and Eric Lipton: "After the boom in natural gas."

39. Ibid.

40. Ibid.

41. The Economist: "America's cheap gas."

42. Ibid.

43. Andrew Maykuth: "Natural gas producers turn to 'green completion.'"

44. Jeannie Kever, FuelFix: "Technology draws bead on hydraulic fracturing," Jan 14, 2013. http://fuelfix.com/blog/2013/01/14/ technology-draws-bead-on-hydraulic-fracturing/. Alison Sider, Russell Gold, and Ben Lefebvre, Wall Street Journal: "Drillers begin reusing 'frack water,'" Nov 20, 2012.

45. Alison Sider, Russell Gold, and Ben Lefebvre: "Drillers begin reusing 'frack water.'"

46. Erica Geis, New York Times: "Race is on to clean up hydraulic fracturing," Dec 4, 2012.

47. Ibid.

48. Ibid.

49. Ibid.

50. Basetrace.com: "About our technology." http://www.basetrace. com/technology.

51. Cecelia Mason, West Virginia Public Broadcasting: "SkyTruth expands, offers Marcellus drilling data," Oct 27, 2011. http:// www.wvpubcast.org/newsarticle.aspx?id=22364.

52. Andrew Revkin, Dot Earth: "The do-it-yourself approach to tracking gas drilling," Nov 14, 2012. http://dotearth.blogs. nytimes.com/2012/11/14/the-d-i-y-do-it-yourself-approach-to-tracking-frackin/.

53. Andrew C. Revkin, New York Times, Dot Earth: "Can public leak patrols stem gas emissions at a profit?," Nov 16, 2012. http://dote-arth.blogs.nytimes.com/2012/11/16/can-public-leak-patrols-s tem-gas-emissions-at-a-profit/.

54. Ibid.

55. Leslie Kaufman, New York Times, Green Blog: "Fracking and water: E.P.A. zeroes in on 7 sites," Jun 23, 2011. http:// green.blogs.nytimes.com/2011/06/23/fracking-and-water-e-p-a-zeroes-in-on-7-sites.

56. Tenille Tracy, RigZone, DJ Newswire: "Chesapeake Energy to host EPA in study of fracking risk to water," Jan 24, 2013. http://www. rigzone.com/news/oil_gas/a/123748/Chesapeake_Energy_to_ Host_EPA_in_Study_of_Fracking_Risk_to_Water.

57. EPA: "Potential impacts of hydraulic fracturing on drinking water resources: Progress report." http://www2.epa.gov/sites/produc-tion/files/documents/hf-report20121214.pdf#page=18.

58. Wall Street Journal: "Chesapeake to host EPA in study of fracking risk: Natural-gas producer to host federal study of drilling's risk to water quality," Jan 23, 2013. http://online.wsj.com/article/SB1 000142412788732385490457826017309918476.html.

59. Elisabeth Rosenthal, New York Times: "Life after oil and gas," Mar 23, 2013. http://www.nytimes.com/2013/03/24/sunday-review/life-after-oil-and-gas.html.

60. Andrea Bernstein, WNYC: "Obama: Cut oil imports by a third in the next decade," Mar 30, 2011. http://www.wnyc.org/blogs/transportation-nation/2011/mar/30/obama-cut-oil-consumption-by-a-third-in-the-next-decade.

61. Andrew E. Kramer, New York Times: "Russia skips hybrids in push for cars using natural gas," Apr 11, 2013.

62. Tom Fowler, Wall Street Journal: "America, start your natural-gas engines," Jun 18, 2012. http://online.wsj.com/article/SB10001424052702304192704577406431047638416.html.

63. Ibid. and AAA NewsRoom: "AAA Monthly Gas Price Report: highest Increase in Gas Prices to Begin Year on Record," Feb 28, 2013 newsroom.aaa.com

64. Ibid.

65. Andrew E. Kramer, New York Times: "Russia skips hybrids in push for cars using natural gas," Apr 11, 2013.

66. Tom Fowler: "America, start your natural-gas engines."

67. Elizabeth Miller, New York Times: "China must exploit its shale gas," Apr 12, 2013.

68. Ibid.

69. Ibid.

70. Ibid.

71. Andrew Revkin, New York Times, Opinion, DotEarth Blog: "A fracking method with fewer water woes?" Nov 8, 2011. http://dotearth.blogs.nytimes.com/2011/11/08/a-fracking-method-with-fewe-water-woes/.

Conclusion

1. Ben Sills, Bloomberg.com: "Solar may produce most of world's power by 2060, IEA says," Aug 29, 2011. http://www.bloomberg.

com/news/2011-08-29/solar-may-produce-most-of-worlds-power-by-2060-iea-says.html.

2. Elisabeth Rosenthal, New York Times: "Life after oil and gas," Mar 23, 2013. http://www.nytimes.com/2013/03/24/sunday-review/life-after-oil-and-gas.html.

3. Renewable Energy Policy Network for the 21st Century: "Renewables 2010 Global status report," 2010.

4. Erik Kirschbaum, Reuters: "Renewable energy fosters a boom in depressed German state," Oct 6, 2011.

5. Elisabeth Rosenthal: "Life after oil and gas."

6. Michael McElroy and Xi Lu, Harvard Magazine: "Fracking's future: Natural gas, the economy, and America's energy prospects," Jan–Feb 2013.

7. Euractiv.com: "IEA chief: Energy efficiency directive is 'a must,'" Jun 14, 2012. http://www.euractiv.com/energy-efficiency/fatih-birol-iea-chief-energy-eff-news-513287.

8. Michael McElroy and Xi Lu: "Fracking's future."

FURTHER READING

Hydrofracking is a large and constantly shifting subject, one that is extensively covered by the mainstream press. For readers interested in general references, I suggest consulting the Notes chapter in this book. For those looking to pursue a particular line of thought or technical detail, or to dig deeper into complementary aspects of hydrofracking (such as energy use or climate science), below is a list of specialty publications, government reports, academic and industry studies, blogs, and the like. This is just a starting point, of course, but it should provide a useful pathway to understanding these timely issues.

American Petroleum Institute (API): http://www.api.org

American Exploration and Production Council (AXPC): http://www.axpc.us

America's Natural Gas Alliance (ANGA): http://anga.us

Bakken Shale news, marketplace, jobs: http://bakkenshale.com

Barnett Shale Energy Education Council: http://www.bseec.org

British Petroleum (BP): http://www.bp.com/content/dam/bp/pdf/investors/6423_BP_Unconventional_Gas.pdf

Bureau of Land Management (BLM): http://www.blm.gov/wo/st/en.html

California Energy Commission: http://www.energy.ca.gov

Ceres: http://www.ceres.org

Chesapeake Energy: http://www.chk.com

Council on Foreign Relations: http://www.cfr.org

US Department of Energy: http://energy.gov

US Department of the Interior: http://www.doi.gov/index.cfm

Earth Justice: http://earthjustice.orgceres

Energy in Depth: http://energyindepth.org

US Energy Information Administration (EIA): http://www.eia.gov

US Environmental Protection Agency: http://www2.epa.gov/hydraulicfracturing

FracFocus: http://fracfocus.org

Food & Water Watch: http://www.foodandwaterwatch.org

GasLand (HBO): http://www.hbo.com/documentaries/gasland/index.html

US Geological Survey (USGS): http://energy.usgs.gov/OilGas/UnconventionalOilGas/HydraulicFracturing.aspx

Geology.com: http://geology.com

US House of Representatives, Energy and Commerce Committee: http://energycommerce.house.gov

Levi, Michael A., *The Power Surge: Energy, Opportunity, and the Battle for America's Future*, a CFR Book, New York: Oxford University Press, May 2013

International Energy Agency (IEA): http://www.iea.org

Marcellus Shale Coalition: http://marcelluscoalition.org

Massachusetts Institute of Technology (MIT) Energy Initiative: http://mitei.mit.edu

The National Academies (National Academy of Sciences, National Academy of Engineering, Institute of Medicine, National Research Council): http://www.nas.edu

National Petroleum Council: http://npc.org

National Renewable Energy Laboratory: http://www.nrel.gov

NaturalGas.org http://www.naturalgas.org

Natural Resources Defense Council: http://www.nrdc.org

New York City Department of Environmental Protection: http://www. nyc.gov/html/dep/html/home/home.shtml

New York State Department of Environmental Conservation: http:// www.dec.ny.gov

Pacific Institute: http://www.pacinst.org

Pennsylvania Department of Environmental Protection: http://www. depweb.state.pa.us/portal/server.pt/community/dep_home/5968

Pickens Plan: http://www.pickensplan.com

ProPublica: http://www.propublica.org/series/fracking

The Railroad Commission of Texas: http://www.rrc.state.tx.us

Royal Dutch Shell: http://www.shell.com

Strategic Unconventional Fuels Task Force: http://www.unconventionalfuels.org/home.html

350.org: http://350.org

Yergin, Daniel: http://danielyergin.com. And see: *The Quest: Energy, Security, and the Remaking of the Modern World*, New York: Penguin, September 2011.

INDEX

Printed in the USA/Agawam, MA
February 3, 2014

584865.180